This is your very own math book! You can write in it, draw, circle, and color as you explore the exciting world of math.

Let's get started. Grab a crayon and draw a picture that shows what math means to you.

Have fun!

This is your space to draw.

Education

Bothell, WA • Chicago, IL • Columbus, OH • New York, NY

connectED.mcgraw-hill.com

STEM McGraw-Hill is committed to providing instructional materials in Science, Technology, Engineering, and Mathematics (STEM) that give all students a solid foundation, one that prepares them for college and careers in the 21st century.

Send all inquiries to:
McGraw-Hill Education
STEM Learning Solutions Center
8787 Orion Place
Columbus, OH 43240

ISBN: 978-0-02-116068-6 *(Volume 2)*
MHID: 0-02-116068-6

Printed in the United States of America.

9 10 DOR 19 18 17 16 15 14 13

Our mission is to provide educational resources that enable students to become the problem solvers of the 21st century and inspire them to explore careers within Science, Technology, Engineering, and Mathematics (STEM) related fields.

Meet The Artists!

Wilmer Cortez Cabrera

Numbers in my Life When we heard that I was a winner, my class friends hugged me so much that I fell on the floor. I feel like I am a star. *Volume 1*

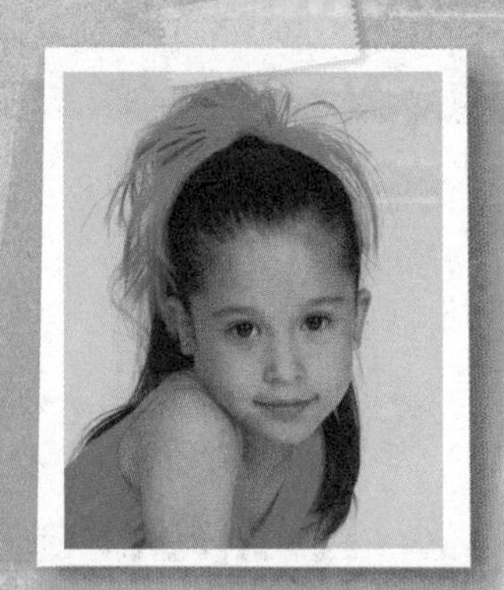

Samantha Garza

I Add and Subtract I like to read, dance and play. Making this art work was fun. *Volume 2*

Other Finalists

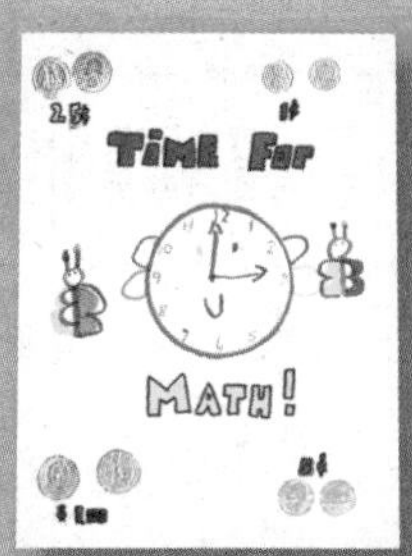

K. Jock's and M. Kennedy's Class*
Time and Money is Math

Carly Gordon
Math and Art go great together!

Manuel Otero
Line Math

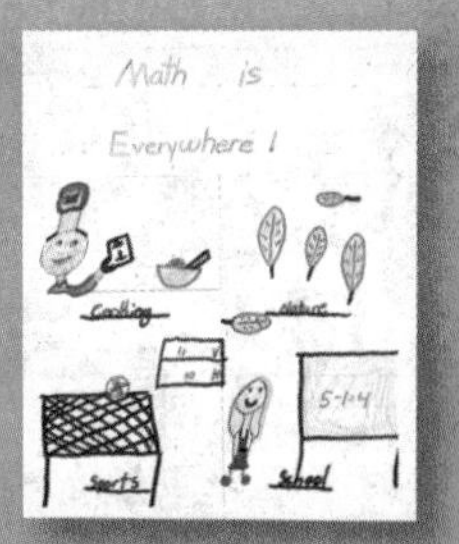

Katy Rupnow
Math is Everywhere!

Ma Myat Thiri Kyaw
Math Swamp

Jahni Williams
All About Numbers

Nora Carter's Class
Math for Life

Brittany Schweitzer
Serving up Math

Lillian Gaggin
Wristwatch

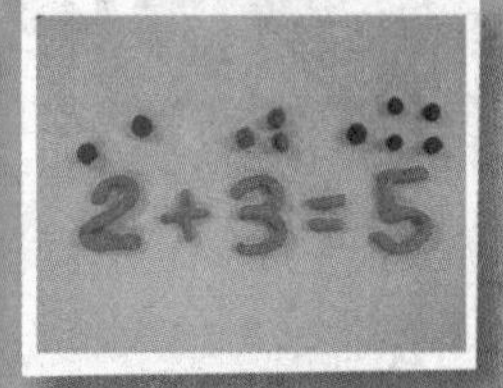

Kristie Mendez's Class*
Add Up the Dough

Find out more about the winners and other finalists at www.MHEonline.com.

We wish to congratulate all of the entries in the 2011 *McGraw-Hill My Math* "What Math Means To Me" cover art contest. With over 2,400 entries and more than 20,000 community votes cast, the names mentioned above represent the two winners and ten finalists for this grade.

** Please visit mhmymath.com for a complete list of students who contributed to this artwork.*

it's all at
connectED.mcgraw-hill.com

Go to the Student Center for your eBook, Resources, Homework, and Messages.

Write your Username

Password

Get your resources online to help you in class and at home.

Vocab

Find activities for building vocabulary.

Watch

Watch animations of key concepts.

Tools

Explore concepts with virtual manipulatives.

Check

Self-assess your progress.

eHelp

Get targeted homework help.

Games

Reinforce with games and apps.

Tutor

See a teacher illustrate examples and problems.

GO mobile

Scan this QR code with your smart phone* or visit mheonline.com/stem_apps.

*May require quick response code reader app.

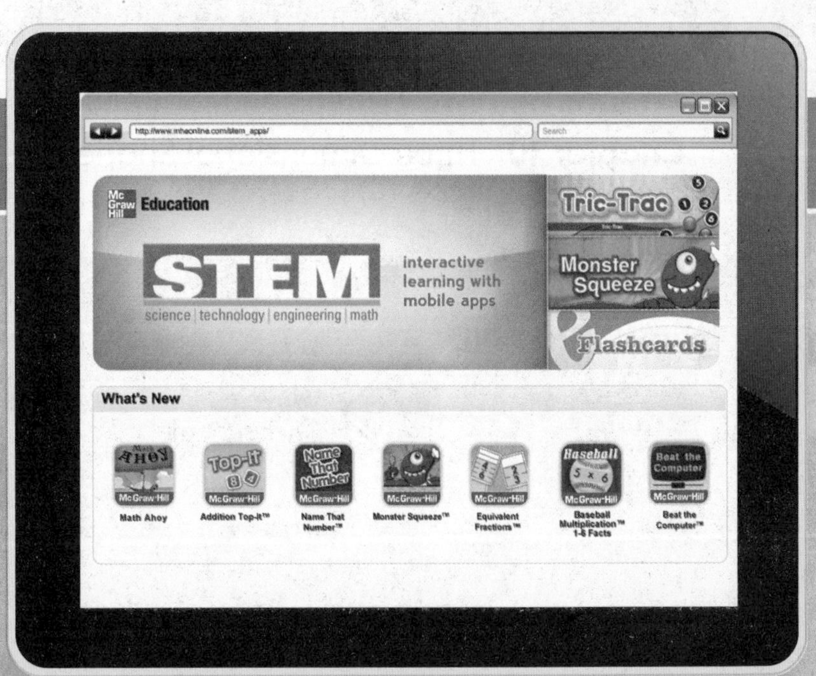

Contents in Brief

Organized by Domain

The suggested pacing supports 1 day per lesson for instruction, 2 days per chapter for review and assessment, and includes additional time for remediation and differentiation.

Chapter 1 Addition Concepts

ESSENTIAL QUESTION
How do you add numbers?

Getting Started

Lessons and Homework

Wrap Up

connectED.mcgraw-hill.com

Chapter 2 Subtraction Concepts

ESSENTIAL QUESTION
How do you subtract numbers?

Getting Started

Lessons and Homework

Wrap Up

connectED.mcgraw-hill.com

Your safari adventure starts online!

Chapter 3 Addition Strategies to 20

ESSENTIAL QUESTION
How do I use strategies to add numbers?

Getting Started

Let's roll into the big city!

Lessons and Homework

Wrap Up

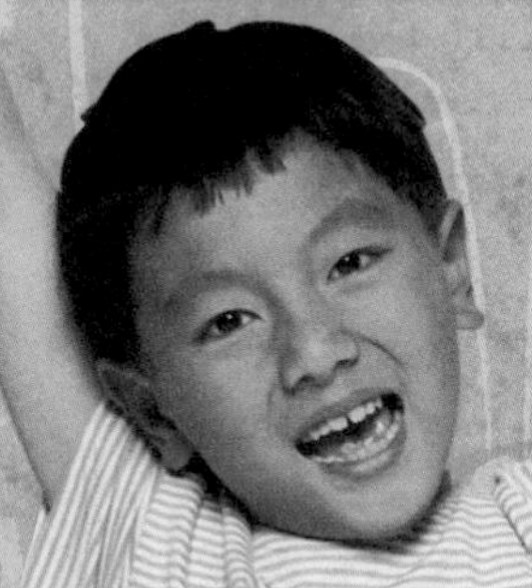

Look for this!
Click online and you can watch videos that will help you learn the lessons.

connectED.mcgraw-hill.com

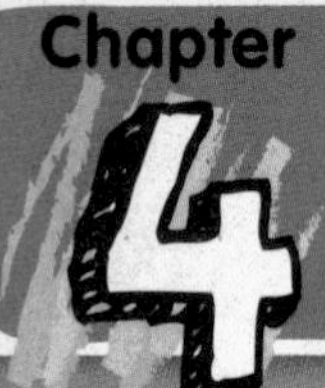

Chapter 4 Subtraction Strategies to 20

ESSENTIAL QUESTION
What strategies can I use to subtract?

Getting Started

Lessons and Homework

I love the beach!

Wrap Up

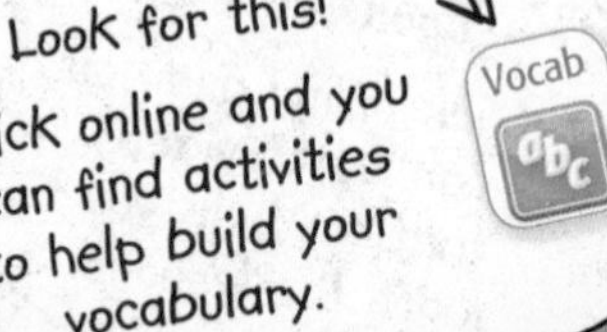

connectED.mcgraw-hill.com

Chapter 5 Place Value

ESSENTIAL QUESTION
How can I use place value?

Getting Started

Lessons and Homework

We're going to the toy store!

Wrap Up

There are fun games online!

Chapter 6 Two-Digit Addition and Subtraction

ESSENTIAL QUESTION
How can I add and subtract two-digit numbers?

Getting Started

Lessons and Homework

Wrap Up

You can find fun activities online!

connectED.mcgraw-hill.com

Chapter 7 Organize and Use Graphs

ESSENTIAL QUESTION
How do I make and read graphs?

Getting Started

Let's get active!

Lessons and Homework

Wrap Up

Look for this!
Click online and you can find tools that will help you explore concepts.

connectED.mcgraw-hill.com

Chapter 8 Measurement and Time

ESSENTIAL QUESTION
How do I determine length and time?

Getting Started

Lessons and Homework

Look! I'm a watch dog!

Wrap Up

My classroom is fun!

connectED.mcgraw-hill.com

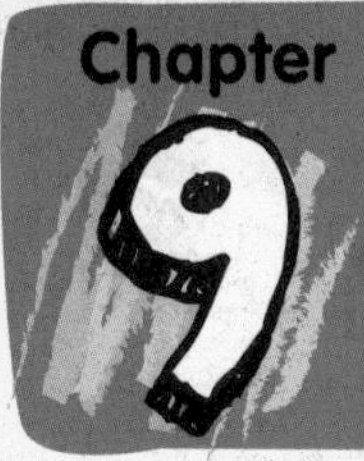

Chapter 9 Two-Dimensional Shapes and Equal Shares

ESSENTIAL QUESTION
How can I recognize two-dimensional shapes and equal shares?

Getting Started

Lessons and Homework

We're going to the farm!

Wrap Up

Look for this!
Click online and you can check your progress.

connectED.mcgraw-hill.com

Chapter 10 Three-Dimensional Shapes

ESSENTIAL QUESTION
How can I identify three-dimensional shapes?

Getting Started

Lessons and Homework

Wrap Up

Look for this!
Click online and you can get more help while doing your homework.
eHelp

connectED.mcgraw-hill.com

Chapter 7 Organize and Use Graphs

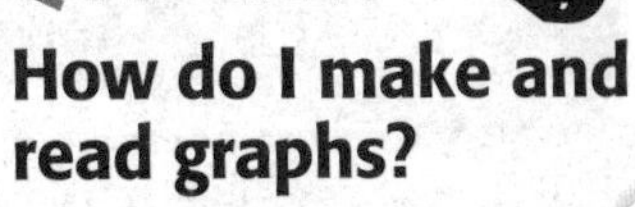

How do I make and read graphs?

We're Getting Fit!

Watch a video!

Watch

My Common Core State Standards

Measurement and Data

CCSS

1.MD.4 Organize, represent, and interpret data with up to three categories; ask and answer questions about the total number of data points, how many in each category, and how many more or less are in one category than in another.

Standards for Mathematical PRACTICE

1. Make sense of problems and persevere in solving them.
2. Reason abstractly and quantitatively.
3. Construct viable arguments and critique the reasoning of others.
4. Model with mathematics.
5. Use appropriate tools strategically.
6. Attend to precision.
7. Look for and make use of structure.
8. Look for and express regularity in repeated reasoning.

= focused on in this chapter

Name

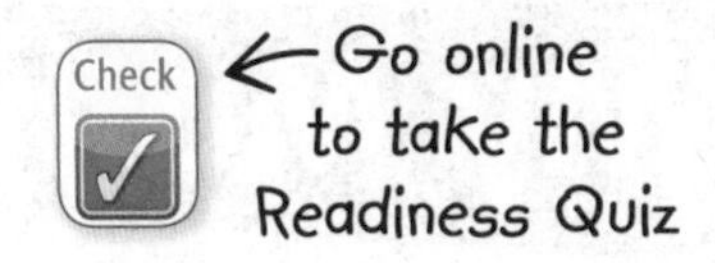

Count. Write the number of objects you counted.

1.  3

2.  7

Circle the correct answer.

3.  is more than / is fewer than 

4.  is more than /  is fewer than 

Circle the correct answer.

5. 3 ducks are in the pond. 2 ducks are in the barn. Which place has more ducks?

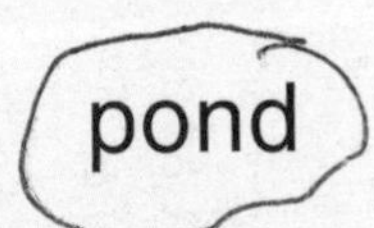

barn

How Did I Do? Shade the boxes to show the problems you answered correctly.

1	2	3	4	5

Name

My Math Words

Review Vocabulary

count shape size

Use the words to write how to sort each group.

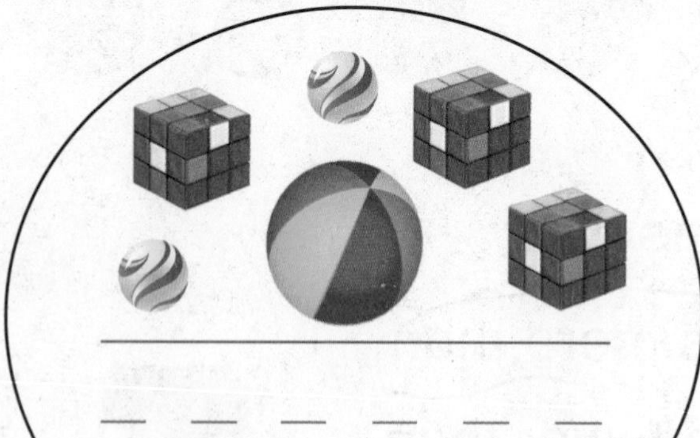

Sort by:

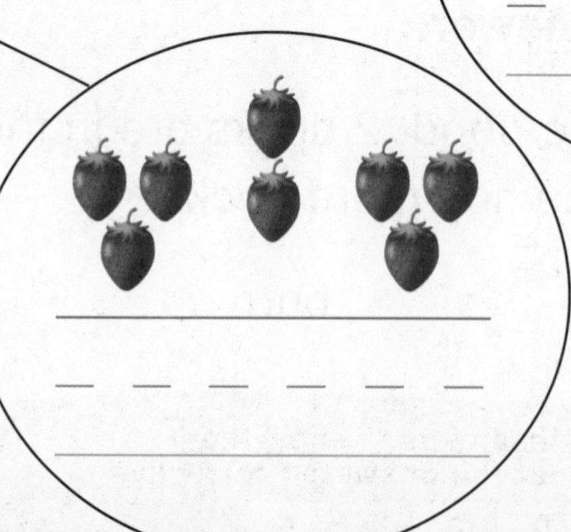

My Vocabulary Cards

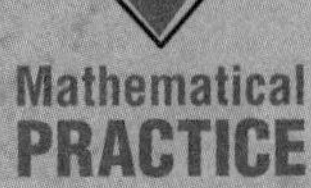

Lesson 7-5

bar graph

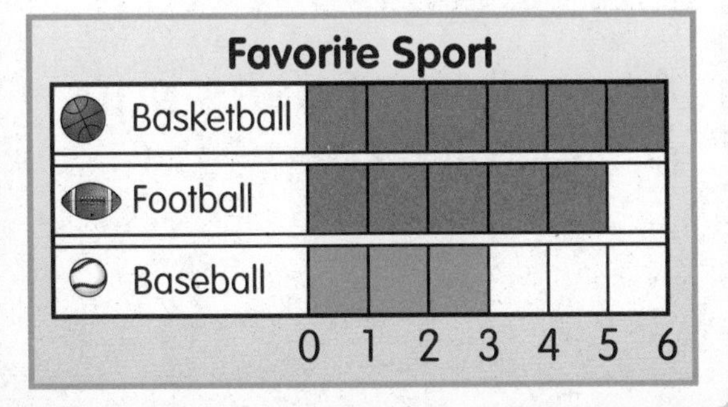

Lesson 7-3

data

Favorite Sport					
Baseball					
Soccer					
Basketball					

Lesson 7-3

graph

Fruit We Ate Today

Grapes | Strawberries | Oranges

Lesson 7-3

picture graph

Favorite Crayon Color			
Red			
Blue			
Green			

Lesson 7-1

survey

Lunch Today?	
Pack	𝍸 𝍸
Buy	𝍸 𝍷𝍷𝍷
Buy milk only	𝍷𝍷𝍷𝍷

Lesson 7-1

tally chart

Favorite Snack		
Snack	Tally	Total
Popcorn	𝍸 𝍷𝍷𝍷	8
Berry	𝍷𝍷	2
Pretzel	𝍷𝍷𝍷𝍷	4

Teacher Directions:
Ideas for Use

- Have students write a tally mark on each card every time they read the word in this chapter or use it in their writing.
- Have students sort the vocabulary words by the number of syllables in each word.

Numbers or pictures collected to show information.

A graph that uses bars to show data.

A graph that has different pictures to show data.

A way to present data collected. Also, a type of chart.

A chart that shows a mark for each vote in a survey.

To collect data by asking people the same question.

My Foldable

FOLDABLES® Follow the steps on the back to make your Foldable.

Tally Chart

Favorite Snack		
Snack	Tally	Total
Apple		
Pretzel		
Juice		

Picture Graph

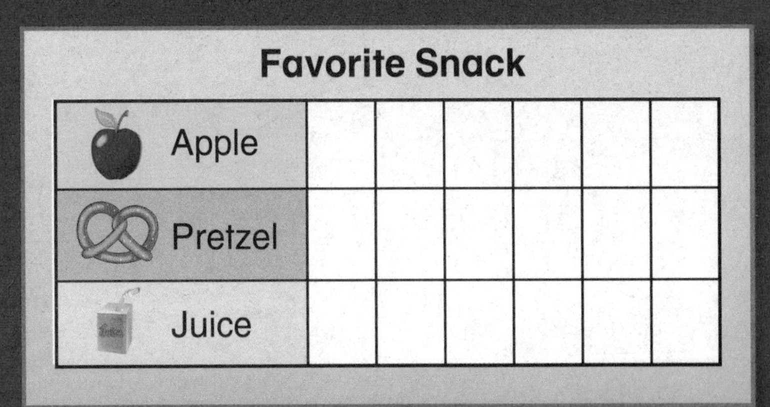

Favorite Snack						
Apple						
Pretzel						
Juice						

Bar Graph

Favorite Snack						
Apple						
Pretzel						
Juice						

0 1 2 3 4 5 6

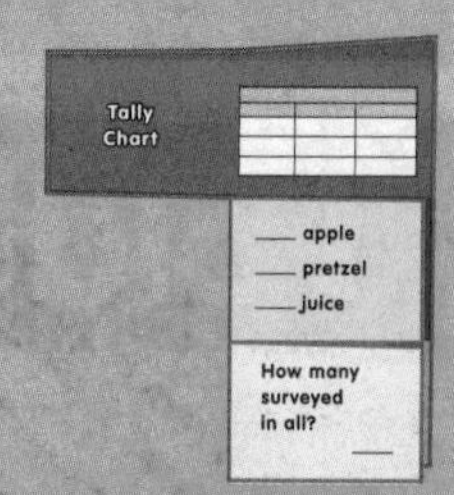

What's Your Favorite Snack?

_____ apple

_____ pretzel

_____ juice

How many surveyed in all?

_____

Name ....

Tally Charts

Lesson 1

ESSENTIAL QUESTION
How do I make and read graphs?

Explore and Explain

My Favorite Activity

Activity	Tally	Total
Run	||	
Dance	||||	
Play outside	|||	

_____ − _____ = _____ more people

Write your subtraction sentence here.

Teacher Directions: Use ■ to show the number of people who voted for each activity. Write each total. How many more people like to dance than to run? Write a subtraction number sentence to solve.

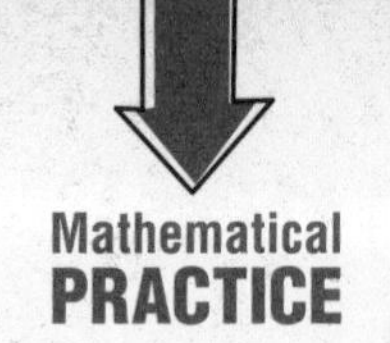

See and Show

A **tally chart** shows a mark for each vote in a survey. A **survey** asks people the same question.

Favorite Vegetables		
Vegetable	Tally	Total
Carrot	|||	3
Pea	||	2
Corn	𝍸 ||	7

| means 1 vote. 𝍸 means 5 votes.

Ask 10 friends to choose their favorite school subject. Make a tally chart. Write the totals.

Favorite Subject		
Subject	Tally	Total
Math		
Reading		
Science		

Use the tally chart. How many chose each subject?

1. MATH ______ **2.** READING ______ **3.** SCIENCE ______

Talk Math How are tally marks used to take surveys?

CCSS

Name ______________________

On My Own

Write the totals. Use the chart to answer the questions.

What is Your Favorite Color?		
Color	Tally	Total
Red	~~IIII~~ III	
Blue	III	
Purple	~~IIII~~	

4. How many people chose red? ______________

5. How many people chose purple? ______________

6. Do more people like purple or blue? ______________

7. How many more people like red than blue?

____

8. Do more people like red or purple? ______________

9. How many fewer people like blue than purple?

____

10. How many people were surveyed in all?

____

Problem Solving

11. Circle the tally chart that shows 2 students like crackers, 6 students like bananas, and 4 students like carrots.

Favorite Snack	
Snack	Tally
Crackers	||
Bananas	~~||||~~ |
Carrots	||||

Favorite Snack	
Snack	Tally
Crackers	||||
Bananas	||
Carrots	~~||||~~ |

HOT Problem

Samantha is having a pizza party.She asks her guests to pick their favorite kind of pizza. If she orders one kind, which one does she order? Explain.

Favorite Pizza Topping		
Topping	Tally	Total
Cheese	~~||||~~ ||	7
Pepperoni	|||	3
Sausage	||	2

Name ....

My Homework

Lesson 1
Tally Charts

Homework Helper

Need help? connectED.mcgraw-hill.com

A tally chart shows a mark for each vote in a survey.
| means 1 vote. 𝍸 means 5 votes.

Favorite Riding Toy		
Riding Toy	Tally	Total
Scooter	||	2
Bicycle	𝍸 ||	7
Skateboard	𝍸	5

How many votes did scooter get?

2 votes

Practice

Write the totals. Use the chart to answer the questions.

Favorite Activity		
Activity	Tally	Total
Art	𝍸 |	
Music	𝍸 |||	
Sports	𝍸	

1. How many more votes did music get than art? ______

2. How many people were surveyed? ______

Write the totals. Use the chart to answer the questions.

My Favorite Season		
Season	Tally	Total
Summer	卌 ||	
Fall	卌 ||||	
Spring	||||	

3. How many people like fall? ___________

4. How many more people like summer than spring? ___________

5. Do 7 people like summer or fall? ___________

6. How many people were surveyed? ___________

Vocabulary Check

Complete each sentence.

tally chart **survey**

7. You can collect data by taking a ___________.

8. A ___________ shows data using tally marks.

Math at Home Ask your child to make a tally chart to show which sport your family likes better: football or baseball.

Name

Problem Solving

STRATEGY: Make a Table

Lesson 2

ESSENTIAL QUESTION
How do I make and read graphs?

Kimi buys a T-shirt. It has a stripe on each sleeve. It has 4 words and a picture on it. Which shirt did she buy?

1 Understand

Underline what you know.
Circle what you need to find.

2 Plan

How will I solve the problem?

3 Solve

I will make a table.

Shirt	Picture	Words	Stripe
1	No	1	Yes
2	Yes	4	Yes
3	Yes	2	No

4 Check

Is my answer reasonable? Explain.

Practice the Strategy

A first grade class collects 11 cans.
A second grade class collects 5 cans.
How many more cans did the first grade class collect?

1 Understand Underline what you know.
Circle what you need to find.

2 Plan How will I solve the problem?

3 Solve I will...

Grade	Cans Collected
1	
2	

____ – ____ = ____ more cans

4 Check Is my answer reasonable? Explain.

Name

Apply the Strategy

1. Count the animals. Make a table.

Animal	How Many?
Chicken	
Dog	
Cow	

Use the table to answer the questions.

2. How many more cows than dogs are there? ______

3. How many animals are there in all? ______

4. Raul, Sophia, and Leah each have a pet. The pets are a bird, a snake, and a cat. Sophia's pet has 2 legs. Raul's pet has 0 legs. Leah's pet has 4 legs.

Whose Pet is a Bird?		
Name	Number of Legs	Pet
Raul		
Sophia		
Leah		

Whose pet is a bird? ______

Review the Strategies

Choose a strategy

- Make a table.
- Write a number sentence.
- Draw a diagram.

5. There are 3 plates. Each plate has 2 carrots. How many carrots are there in all?

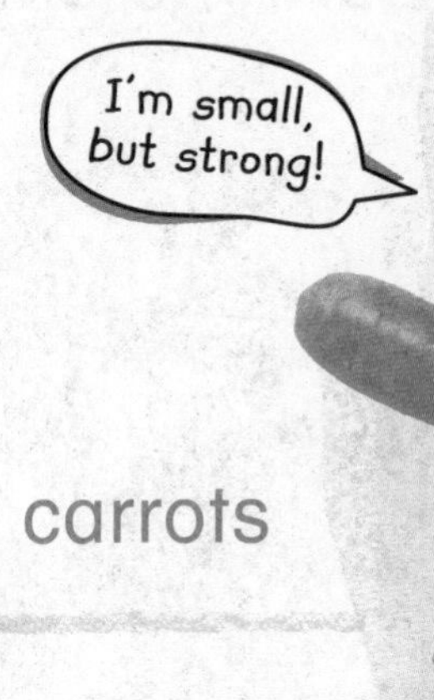

_____ carrots

6. Owen ate 10 oranges. Quinn ate 2 oranges. How many more oranges did Owen eat than Quinn?

_____ more oranges

7. There are 3 maple trees. There are 5 oak trees. There are 8 redwood trees. How many oak and maple trees are there altogether?

_____ trees

Name ....

My Homework

Lesson 2
Problem Solving: Make a Table

Homework Helper

Need help? connectED.mcgraw-hill.com

Cullen buys a box of cereal. It has a picture of 1 bee and 2 words on it. The box is orange. Which box of cereal does he buy?

1 Understand Underline what you know. Circle what you need to find.

2 Plan How will I solve the problem?

3 Solve I will make a table.

Box	1 Bee Picture	Words	Color
1	Yes	0	Orange
2	Yes	2	Orange
3	No	2	Brown

4 Check Is my answer reasonable?

Problem Solving

Underline what you know. Circle what you need to find. Make a table to solve.

1. Ana's toy has 4 wheels. Corey's toy has 1 wheel. Bryn's toy has 2 wheels. The toys are a unicycle, a bicycle, and a toy car. Who has the bicycle?

Name	Wheels	Riding Toy
Ana		
Bryn		
Corey		

____

2. Cooper buys a pair of shoes. They have black laces, 3 stripes, and are blue. Which shoes does he buy?

Shoes	Black Laces	Stripes	Color
1			
2			
3			

Math at Home Make a table and have your child fill it in using objects in your house. Items could include pets, people, or furniture.

Name ....

Make Picture Graphs

Lesson 3

ESSENTIAL QUESTION
How do I make and read graphs?

Explore and Explain

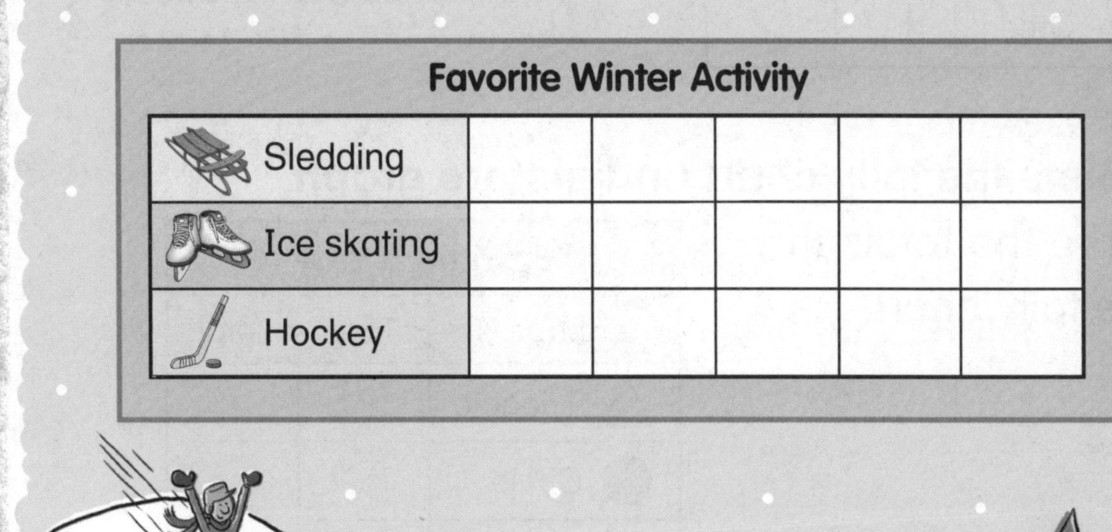

Favorite Winter Activity

Sledding					
Ice skating					
Hockey					

Teacher Directions: Ask 5 students to vote for their favorite winter activity. Show these votes by drawing circles in the boxes.

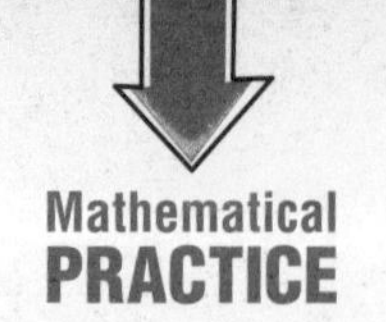

Mathematical PRACTICE

See and Show

A **graph** shows information or **data**.
A **picture graph** uses pictures to show data.
You can use a tally chart to make a picture graph.

Favorite Apple Color		
Color	Tally	Total
Red	|	1
Yellow	|||	3
Green	||	2

Favorite Apple Color			
Red	🍎		
Yellow	🍎	🍎	🍎
Green	🍎	🍎	

Complete the tally chart and picture graph.

1. Write the totals in the tally chart.

Favorite Shape		
Shape	Tally	Total
Triangle	~~||||~~	
Circle	|	
Square	|||	

2. Use the tally chart in Exercise 1 to make a picture graph.

Favorite Shape					
Triangle					
Circle					
Square					

Talk Math What is a picture graph? Describe it.

Name ....

On My Own

Complete the tally chart and picture graph.

3. Write the totals in the tally chart.

Favorite Weather		
Weather	Tally	Total
Sunny	𝍷𝍷	
Rainy	𝍸 𝍷	
Cloudy	𝍸	

4. Use the tally chart in Exercise 3 to make a picture graph.

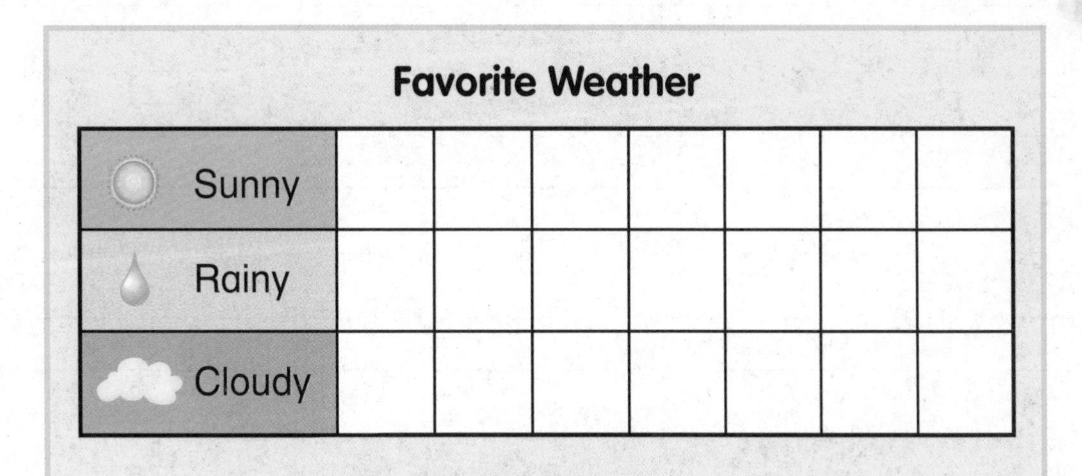

Problem Solving

Mathematical PRACTICE

Complete the picture graph.

5. Miguel asks his friends to name their favorite pet. 3 people like fish. 2 fewer people like cats. 4 people like dogs.

Favorite Pet

Dog						
Cat						
Fish						

Write Math How does a picture show a number on the graph? Explain.

Name ....

My Homework

Lesson 3
Make Picture Graphs

Homework Helper

Need help? connectED.mcgraw-hill.com

A picture graph uses pictures to show data. You can use a tally chart to make a picture graph.

Favorite Flower

Flower	Tally	Total
Daisy	||||	4
Rose	||	2
Tulip	~~||||~~	5

Favorite Flower

Daisy	🌼	🌼	🌼	🌼	
Rose	🌹	🌹			
Tulip	🌷	🌷	🌷	🌷	🌷

Practice

1. Write the totals in the tally chart.

Favorite Juice Flavor

Flavor	Tally	Total
Orange	~~||||~~ |	
Grape	||||	
Lemonade	||	

2. Use the tally chart in Exercise 1 to make a picture graph.

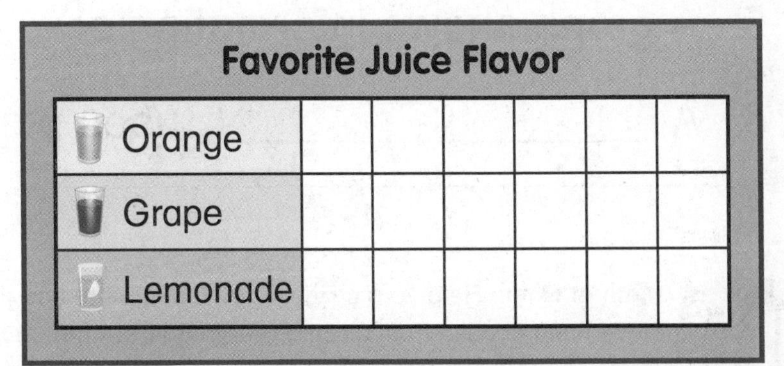

Favorite Juice Flavor

Orange						
Grape						
Lemonade						

3. Write the totals in the tally chart.

Favorite Toy		
Toy	Tally	Total
Bone	IIII	
Rope	~~IIII~~ I	
Ball	~~IIII~~ II	

4. Use the tally chart in Exercise 3 to make a picture graph.

Favorite Toy							
Bone							
Rope							
Ball							

Vocabulary Check 

Complete each sentence.

data **picture graph**

5. A graph shows information or ________________.

6. A ________________ uses pictures to show data.

Math at Home Help your child survey 5 people to find their favorite dinner food. Have them choose from pizza or hamburgers. Then help your child make a picture graph to show this information.

Name ______________________

Read Picture Graphs

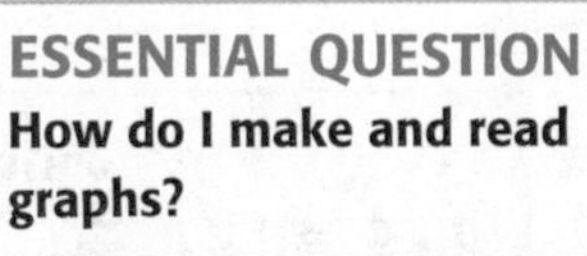

Lesson 4

ESSENTIAL QUESTION
How do I make and read graphs?

Green is good!

Explore and Explain

Favorite Vegetable

Carrot									____
Pea									____
Corn									____

____ people

Teacher Directions: Use ☐ to show the number of people who chose their favorite vegetable. Write how many people like each vegetable. Write how many people were surveyed.

See and Show

The pictures on a picture graph tell how many.

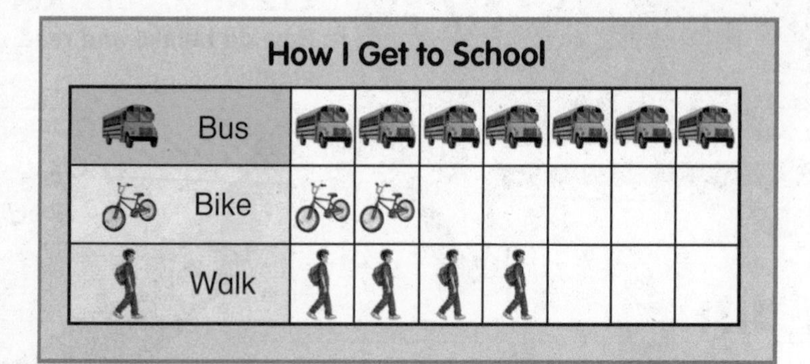

How I Get to School

Bus	🚌	🚌	🚌	🚌	🚌	🚌	🚌
Bike	🚲	🚲					
Walk	🚶	🚶	🚶	🚶			

How many more people ride the bus than walk?

3 people

Use the graph to answer the questions.

Favorite Drink

Chocolate milk	🥛	🥛	🥛	🥛	🥛	🥛	🥛	
Orange juice	🥤	🥤	🥤					
Grape juice	🥤	🥤	🥤					

Mmmmm!

1. How many people were surveyed in all?

____

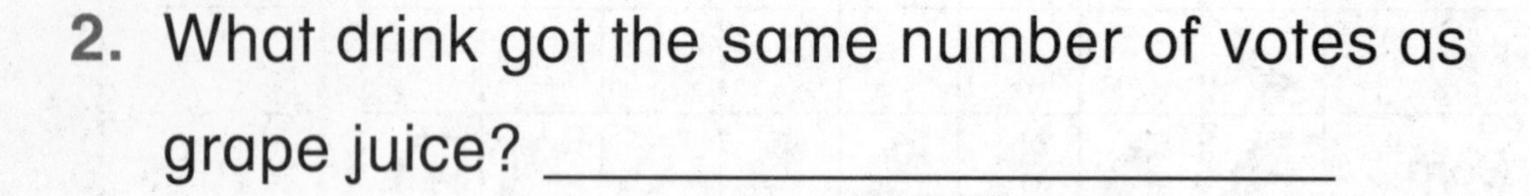

2. What drink got the same number of votes as grape juice? _______________

3. Do more people like chocolate milk or orange juice?

____

Talk Math Explain how you read picture graphs.

Name ....

CCSS

On My Own

Use the graph to answer the questions.

4. How many people chose strawberries?

____

5. Did more people choose oranges or apples?

____

6. Did fewer people choose apples or strawberries?

____

7. How many more people chose strawberries than oranges? _______________

8. How many fewer people chose apples than oranges? _______________

9. How many people were surveyed? _______________

Problem Solving

Mathematical PRACTICE

10. Demarcus is counting sports equipment. He counts 14 pieces of equipment in all. How many jump ropes does he count?

Sports Equipment

Bikes	🚲	🚲	🚲	🚲	🚲	
Jump ropes						
Soccer balls	⚽	⚽	⚽	⚽		

______ jump ropes

HOT Problem Tara asks a question about the picture graph. The answer is the swings. What is the question?

Favorite Playground Equipment

Swings	swing	swing	swing	swing	
Teeter totter	teeter totter	teeter totter			
Slide	slide	slide	slide		

____

____

____

Name ....

My Homework

Lesson 4
Read Picture Graphs

Homework Helper

Need help? connectED.mcgraw-hill.com

The pictures on a picture graph tell how many.

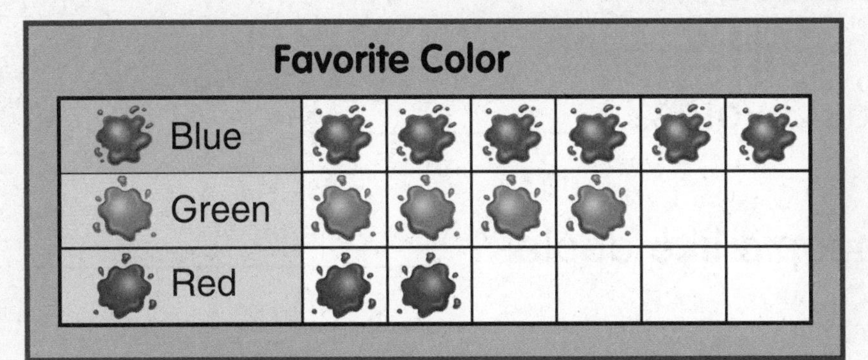

What color has the most votes?

blue

Practice

Use the graph to answer the questions.

1. Do more students like baseball or track?

____

2. How many students voted for basketball?

____

Use the graph to answer the questions.

3. Do more people like bananas or strawberries?

____

4. What fruit has 3 votes? _______________

5. How many people like apples? _______________

6. How many more votes are there for strawberries than for apples? _______________

Test Practice

7. Each picture in a picture graph stands for how many votes?

1	2	3	4
○	○	○	○

Math at Home Have your child gather 3 different kinds of coins from around the house. Have him or her lay them out in separate rows according to the type of coin. Then ask your child questions about the number of coins in each group.

Name

Check My Progress

Vocabulary Check 

Draw lines to match.

1. **picture graph** — A graph that uses different pictures to show data.

2. **data** — To collect data by asking the same question.

3. **survey** — Numbers or pictures collected to show information.

4. **tally chart** — A chart that shows a mark for each vote in a survey.

Concept Check

Write the totals. Use the tally chart to answer the question.

5. Which sport has more votes than basketball?

____

Favorite Sport		
Sport	Tally	Total
Soccer	卌 III	
Basketball	卌	
Baseball	III	

Laura asked her friends to name their favorite color.

Favorite Color		
Color	Tally	Total
Green	||||	
Blue	卌	
Red	||	

Favorite Color					
Green					
Blue					
Red					

6. Write in the totals on the tally chart.

7. Use the tally chart to fill in the picture graph.

8. Did more people choose blue or green?

____

9. How many people chose red? ______________

10. Did fewer people choose blue or red?

____

11. How many more people chose green than red?

____

12. How many people were surveyed? ______________

Name

Make Bar Graphs

Lesson 5

ESSENTIAL QUESTION
How do I make and read graphs?

Fun in the sun!

Explore and Explain

Favorite Summer Activity

Swimming						
Boating						
Water skiing						
0	1	2	3	4	5	6

Teacher Directions: Ask 6 students to choose their favorite summer activity. Have each student shade in 1 box for their activity choice using a crayon. Count each total. Share your totals with a classmate.

See and Show

Mathematical PRACTICE

A **bar graph** uses bars to show information or data. Use the tally chart to make a bar graph.

Favorite Healthy Snack		
Snack	Tally	Total
Apple	ll	2
Cheese	l	1
Celery	lll	3

Favorite Healthy Snack

Apple	■	■	
Cheese	■		
Celery	■	■	■
	0	1 2	3

1. Write the totals in the tally chart.

Shapes I See		
Shape	Tally	Total
Square	𝍸 l	
Triangle	llll	
Circle	𝍸 ll	

2. Use the tally chart in Exercise 1 to make a bar graph.

Shapes I See							
Square							
Triangle							
Circle							

Talk Math What is a bar graph? Describe it.

Name

On My Own

Ask 10 friends to name their favorite playground activity.

3. Write the totals in the tally chart.

Playground Activity		
Activity	Votes	Total
Jump rope		
Slide		
Basketball		

4. Use the tally chart in Exercise 3 to make a bar graph.

Playground Activity									
Jump rope									
Slide									
Basketball									

Problem Solving

5. A survey shows that 1 person likes oatmeal, 8 people like pancakes, and 3 people like scrambled eggs. How many people were surveyed?

_____ people

6. Bryden saw fish and crabs at the beach. He saw 13 animals in all. If he saw 8 crabs, how many fish did he see?

_____ fish

Write Math How do you show each vote on a bar graph? Explain.

Name ....

My Homework

Lesson 5
Make Bar Graphs

Homework Helper

Need help? connectED.mcgraw-hill.com

You can use a tally chart to make a bar graph.

Favorite Way to Travel		
Way to Travel	Tally	Total
Walking	卌 |||	8
Running	卌		6
Skipping	|||	3

Favorite Way to Travel

Walking	■	■	■	■	■	■	■	■
Running	■	■	■	■	■	■		
Skipping	■	■	■					
0	1	2	3	4	5	6	7	8

Practice

1. Write the totals in the tally chart.

Favorite Vehicle		
Vehicle	Tally	Total
Car	卌		
Van	|||	
Truck	卌 ||||	

2. Use the tally chart in Exercise 1 to make a bar graph.

Favorite Vehicle

Car									
Van									
Truck									
0	1	2	3	4	5	6	7	8	9

3. Write the totals in the tally chart.

Favorite Fruit		
Fruit	Tally	Total
Orange	𝍸 |||	
Banana	𝍸 ||	
Cherry	𝍸	

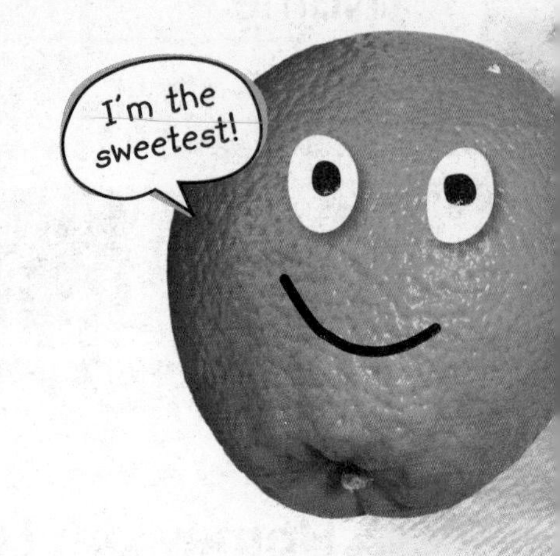

4. Use the tally chart in Exercise 3 to make a bar graph.

Favorite Fruit

Orange									
Banana									
Cherry									
	0	1	2	3	4	5	6	7	8	9

Vocabulary Check

Vocab

Complete each sentence.

bar graph **picture graph**

5. A ________________ uses bars to show data.

6. A ________________ uses pictures to show data.

Math at Home Ask your child to make a bar graph to show the number of pets and the number of people that live in your house.

Name ....

Read Bar Graphs

Lesson 6

ESSENTIAL QUESTION
How do I make and read graphs?

Ready to workout?

Explore and Explain

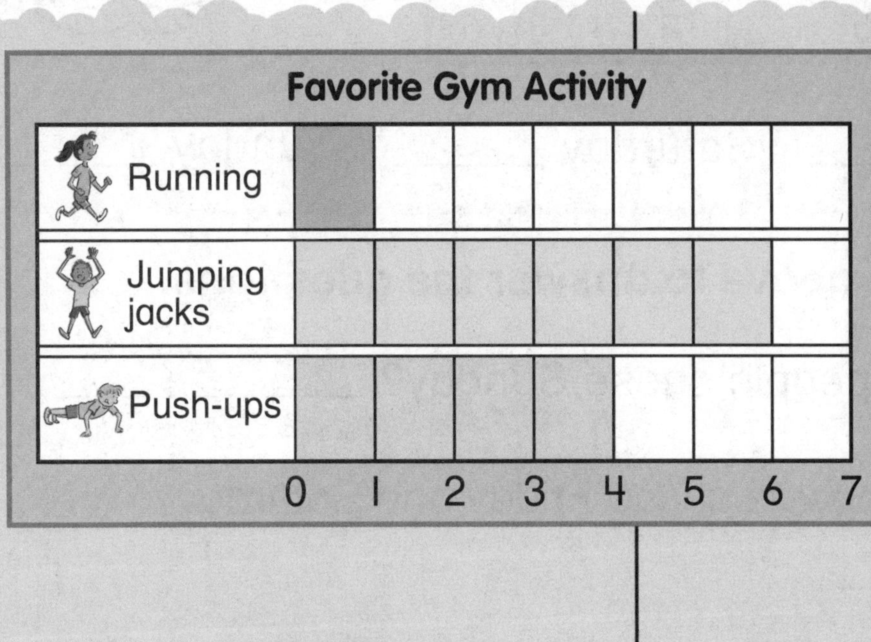

_____

_____

_____

_____ people

Teacher Directions: Use ■ to show the number of people who chose their favorite gym activity. Write how many people chose each gym activity. Write how many people were surveyed in all.

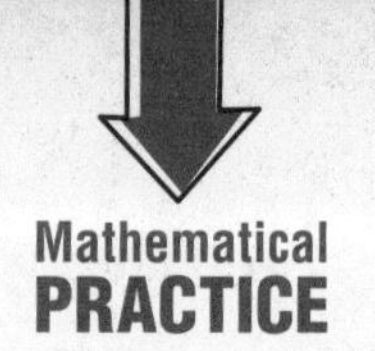

See and Show

Mathematical PRACTICE

The bars on a bar graph tell how many.
Look where each bar ends. Read the number.

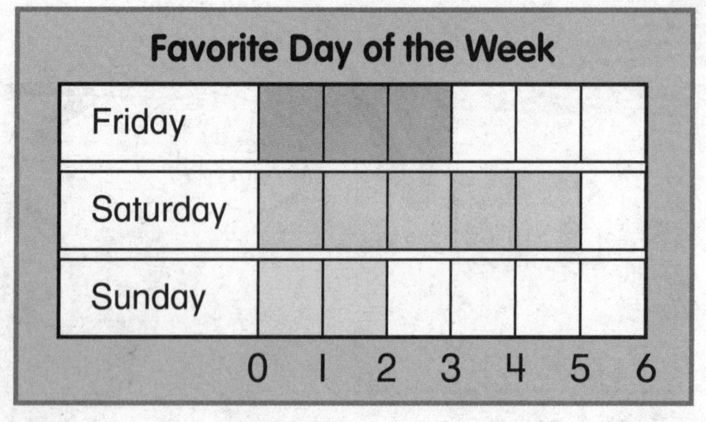

Helpful Hint
The bars on a bar graph can be horizontal or vertical.

Friday __3__ Saturday __5__ Sunday __2__

Use the graph above to answer the questions.

1. How many people chose Sunday? ____________

2. How many people chose Friday and Saturday?

   ____

3. Which day has 1 less vote than Friday?

   ____

4. Which day has 2 more votes than Friday?

   ____

5. How many people were surveyed? ____________

Talk Math Why is the graph above called a bar graph?

Name ____________

On My Own

Use the graph to answer the questions.

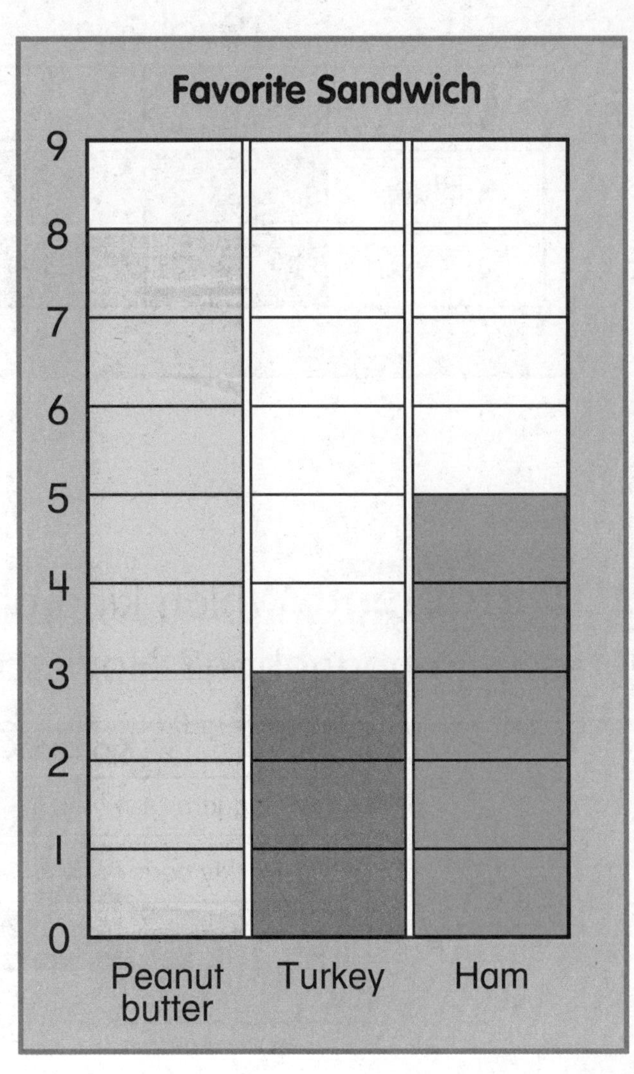

6. How many people chose turkey?

____

7. How many more people chose peanut butter than turkey?

____

8. Did more people choose ham or turkey?

____

9. Which kind of sandwich is liked better than ham?

____

10. Did fewer people choose peanut butter or turkey?

____

11. How many people were surveyed?

____

Problem Solving

12. A class made a bar graph about what color of pencils they used. Which pencil color got less votes than green?

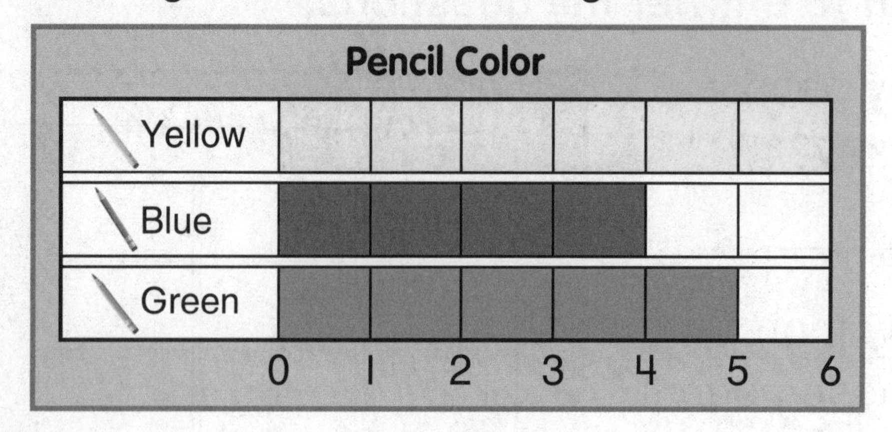

______

HOT Problem Which kind of cacti has more votes than hedgehog? How do you know?

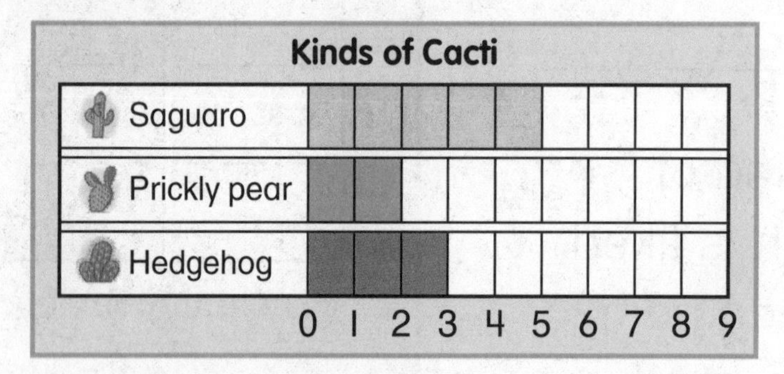

Name ....

My Homework

Lesson 6
Read Bar Graphs

Homework Helper

Need help? connectED.mcgraw-hill.com

The bars on a bar graph tell how many.
Look where each bar ends. Read the number.

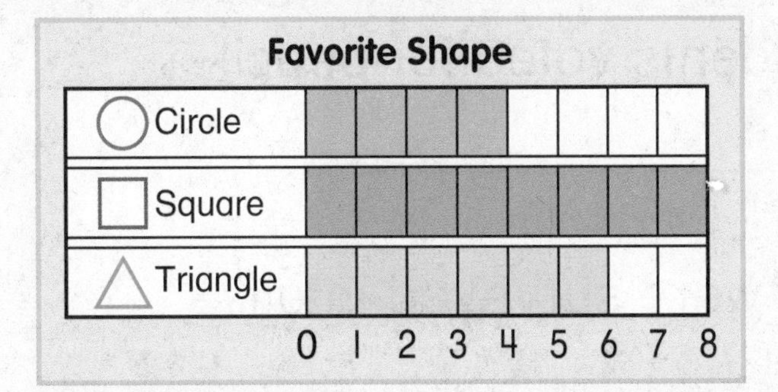

Which shape has the most votes?

square

Practice

Use the graph to answer the questions.

What We Like To Do

Play games
Read a book
Draw
0 1 2 3 4 5 6

1. How many students like to draw? ____________

2. Which activity got fewer votes than playing games?

____

Use the graph to answer the questions.

3. How many students voted for pencil?

____

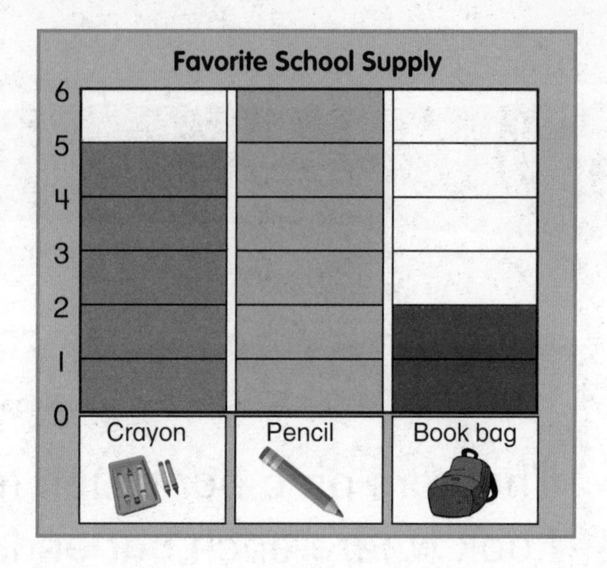

4. Which school supply got fewer than 3 votes?

____

5. How many more students voted for pencil than for book bag? _______________

6. How many students were surveyed in all?

____

Test Practice

7. How many people were surveyed in all?

Favorite Shape

Circle

Square

Triangle

0 1 2 3 4 5 6 7 8

20	18	8	6
○	○	○	○

Math at Home Create a bar graph showing your family members' favorite season. Ask your child questions about this bar graph.

Name ____________________

My Review

Chapter 7

Organize and Use Graphs

Vocabulary Check

Complete each sentence.

bar graph **picture graph**

data **survey**

1. A ____________________ shows data using bars.

2. A ____________________ shows data using pictures.

3. Another word for information is ____________________.

4. You can collect data by taking a ____________________.

Concept Check

Write the totals. Use the chart to answer the question.

5. Do more people like skipping or running?

____

Favorite Way to Move		
Movement	Tally	Total
Walking	|||	
Running	||||	
Skipping	|||	

6. Write the totals in the tally chart.

Favorite Sea Animal		
Animal	Tally	Total
Whales	||	
Dolphins	𝍸 |	
Sharks	𝍸 ||||	

---

7. Use the tally chart in Exercise 6 to make a bar graph.

Favorite Sea Animal									
Whales									
Dolphins									
Sharks									
	0	1	2	3	4	5	6	7	8 9

Use the graphs above to answer the questions.

8. How many people chose dolphins?

____

9. What sea animal got fewer than 5 votes?

____

10. How many people chose sharks and whales?

____

11. How many students were surveyed?

____

Name

Problem Solving

Complete the tally chart.

12. Adam asked his friends about their favorite subjects.

Favorite Subject		
Subject	Tally	Total
Math	𝍸 I	
Science	IIII	
Reading	III	

Test Practice

13. Gemma counts how many glasses of water each person drinks in one day. How many glasses does her dad drink?

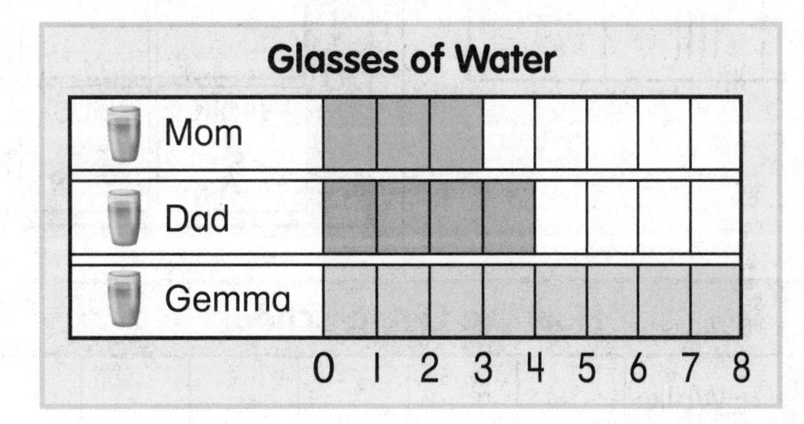

0 ○ 3 ○ 4 ○ 8 ○

Reflect

Chapter 7

Answering the Essential Question

Use the data to make each graph.

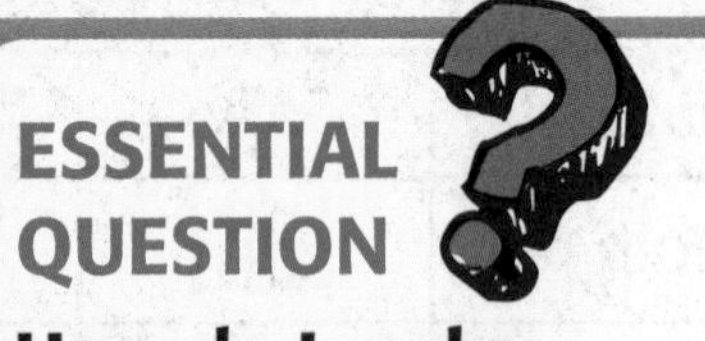

ESSENTIAL QUESTION

How do I make and read graphs?

3 people walk to school.
6 people ride the bus.
4 people ride their bikes.

How We Get to School		
Way	Tally	Total
Walk	|||	
Bus	~~||||~~ |	
Bike	||||	

How We Get to School

	Walk	Bus	Bike
8			
7			
6			
5			
4			
3			
2			
1			
0			

How We Get to School

Walk						
Bus						
Bike						

Now I Know!

Chapter

Measurement and Time

ESSENTIAL QUESTION
How do I determine length and time?

My School Rules!

Watch a video!

Watch

My Common Core State Standards

Measurement and Data

1.MD.1 Order three objects by length; compare the lengths of two objects indirectly by using a third object.

1.MD.2 Express the length of an object as a whole number of length units, by laying multiple copies of a shorter object (the length unit) end to end; understand that the length measurement of an object is the number of same-size length units that span it with no gaps or overlaps.

1.MD.3 Tell and write time in hours and half-hours using analog and digital clocks.

Standards for Mathematical PRACTICE

1. Make sense of problems and persevere in solving them.
2. Reason abstractly and quantitatively.
3. Construct viable arguments and critique the reasoning of others.
4. Model with mathematics.
5. Use appropriate tools strategically.
6. Attend to precision.
7. Look for and make use of structure.
8. Look for and express regularity in repeated reasoning.

= focused on in this chapter

Name

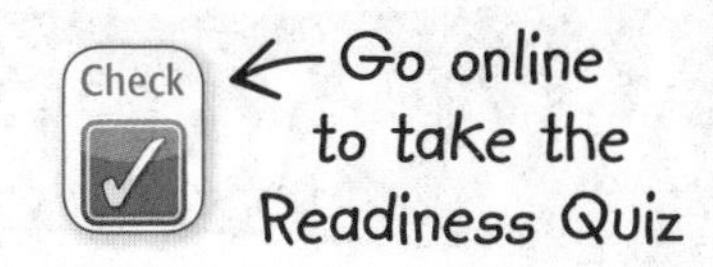

Circle the object that is longer.

1.

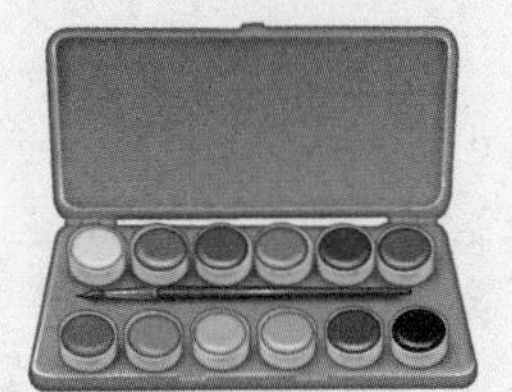

2.

Look at the string.

3. Draw something that is longer.

4. Draw something that is shorter.

5. What number comes next?

6, 7, 8, ______

6. I come just after 10. I am just before 12.
What number am I?

______

How Did I Do?

Shade the boxes to show the problems you answered correctly.

1	2	3	4	5	6

Name

My Math Words

Review Vocabulary

longer shorter

Read each question. Circle the correct answer. In the box at the bottom, draw an object that is shorter than your pencil.

Let's Compare!

Which is longer?

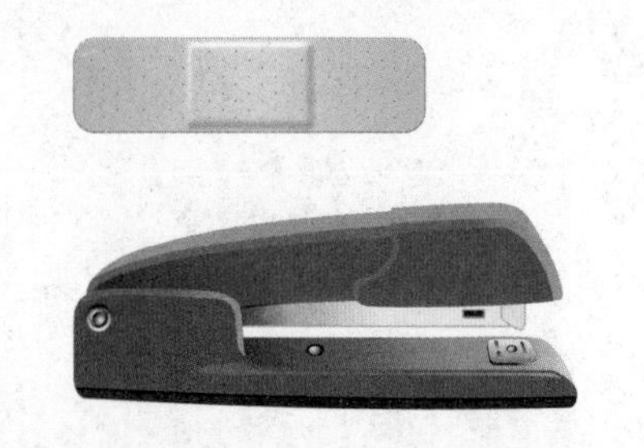

Which is shorter?

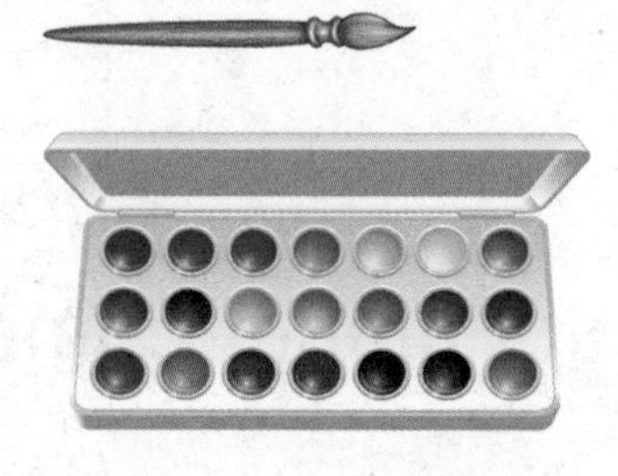

My Drawing!

My Vocabulary Cards

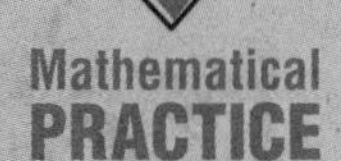

Lesson 8-5

analog clock

Lesson 8-6

digital clock

Lesson 8-7

half hour

half past 5

5:30

Lesson 8-5

hour

1 hour = 60 minutes

Lesson 8-5

hour hand

Lesson 8-1

length

Teacher Directions: Ideas for Use

- Have students group common words. Ask them to add a word that is unrelated to the group. Have them work with a classmate to name the unrelated word.
- Ask students to group like ideas from the chapter. Discuss as a class what strategies they use to help them understand and master these concepts.

A clock that uses only numbers to show time.

A clock that uses an hour hand, a minute hand, and numbers to show time.

A unit to measure time.
1 hour = 60 minutes

One half of an hour is equal to 30 minutes. It is sometimes called *half past* or *half past the hour*.

A measure that tells how long something is.

The hand on an analog clock that tells time to the hour. It is the shorter hand.

My Vocabulary Cards

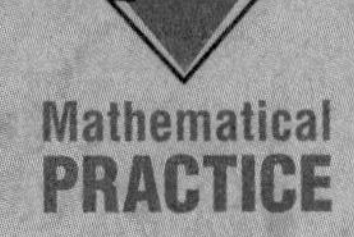

Lesson 8-1

long

long

longer

longest

Lesson 8-3

measure

Lesson 8-5

minute

I minute = 60 seconds

Lesson 8-5

minute hand

Lesson 8-5

o'clock

It is 9 o'clock.

Lesson 8-1

short

short

shorter

shortest

Teacher Directions:
More Ideas for Use

- Ask students to arrange cards to show an opposite pair. Have them explain the meaning of their pairing.
- Ask students to find pictures in books, around the classroom, or in magazines to show each word. Have them work with a friend to guess which word the picture shows.

To find the length of an object using standard or nonstandard units.

A way to compare the lengths of two (or more) objects.

The hand on an analog clock that tells the minutes. It is the longer hand.

A unit to measure time.
1 minute = 60 seconds

A way to describe the length of two (or more) objects.

A word used to tell time. It describes the position of the minute hand at the beginning of the hour.

My Vocabulary Cards

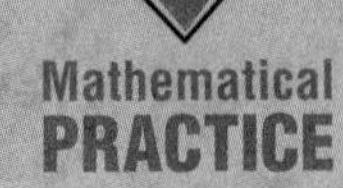

Lesson 8-3

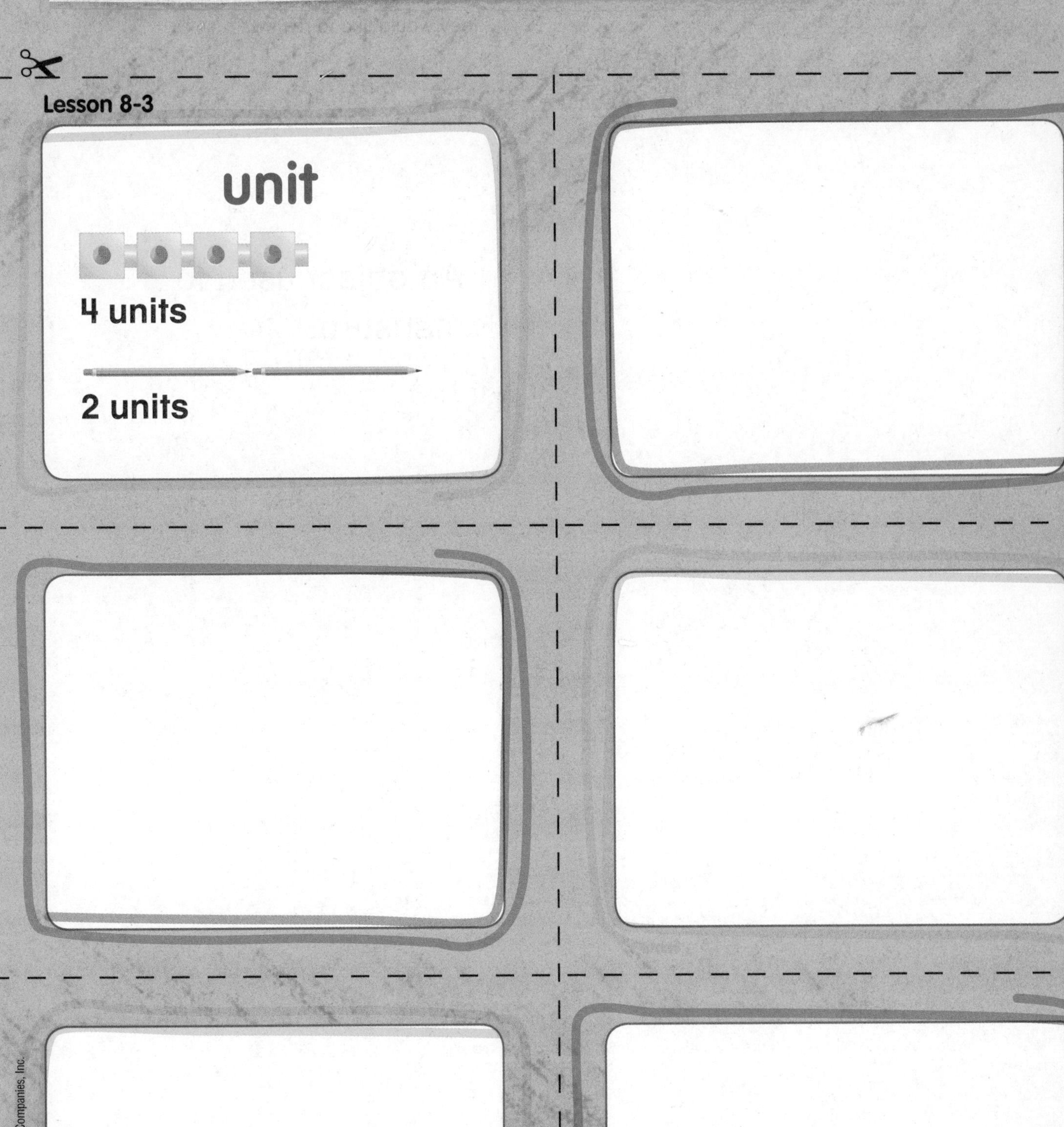

Teacher Directions:
More Ideas for Use

- Ask students to use the blank cards to write their own vocabulary words.
- Have students use the blank cards to write a word from a previous chapter that they would like to review.

An object used to measure.

My Foldable

FOLDABLES® Follow the steps on the back to make your Foldable.

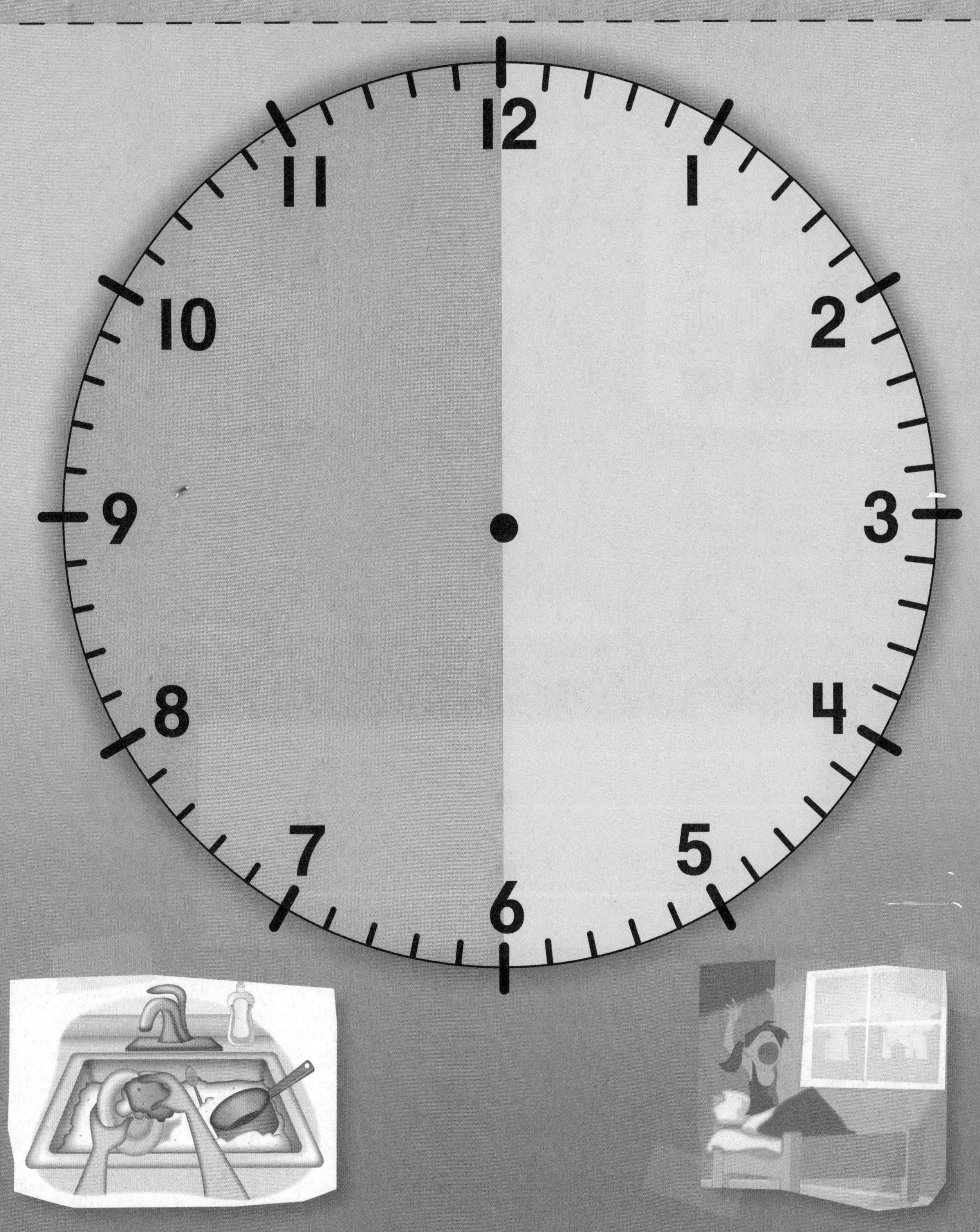

1

2

3

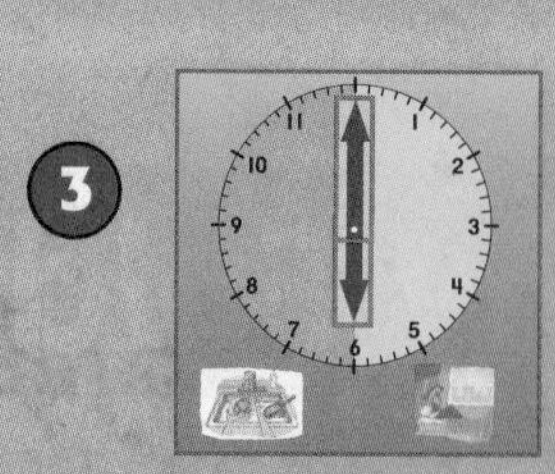

Name ....

Compare Lengths

Lesson 1

ESSENTIAL QUESTION
How do I determine length and time?

Explore and Explain

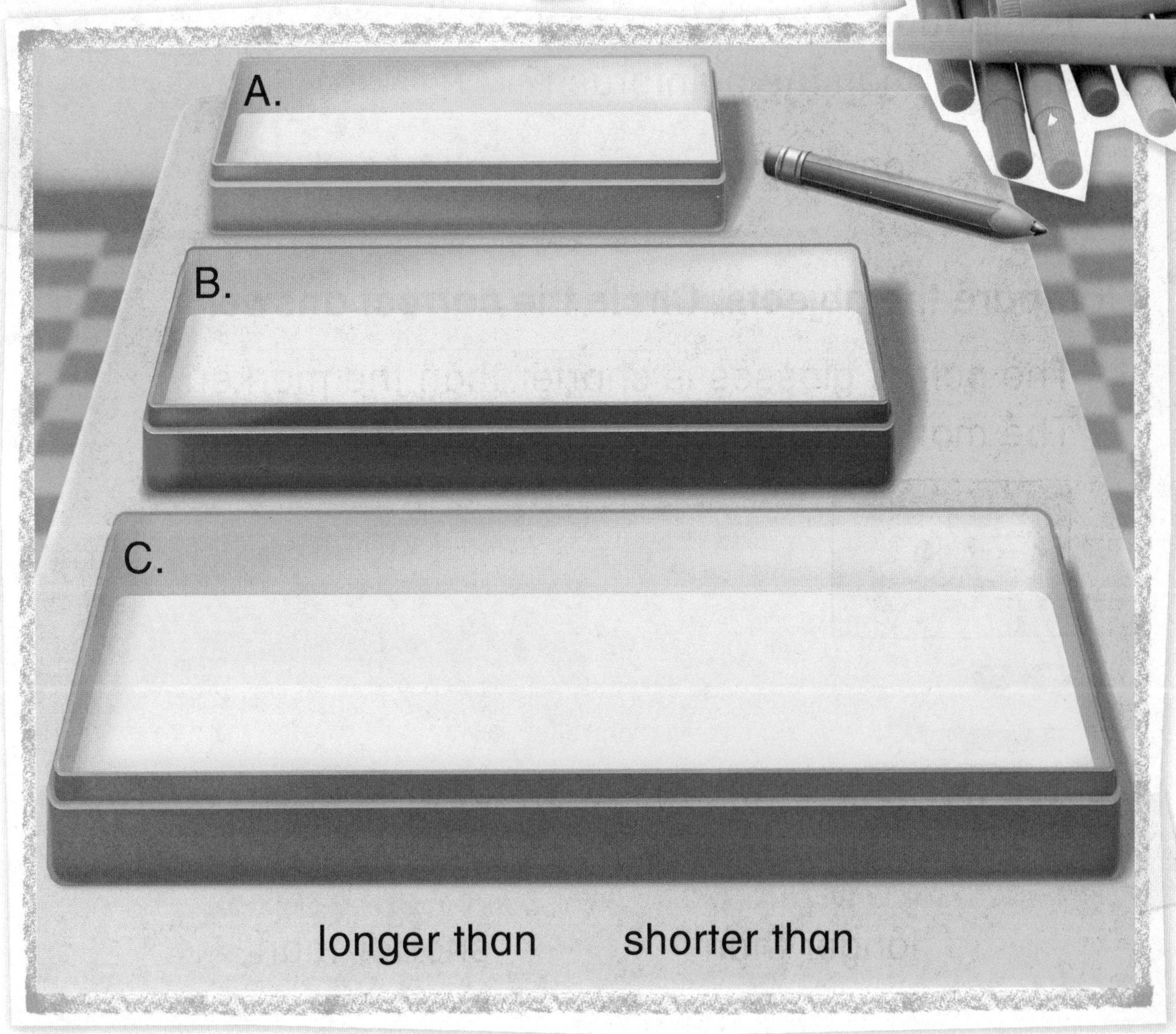

Teacher Directions: Look around the classroom. Choose an object that will fit in each box. Draw the object in the box. Is object A shorter than object B? Is object B shorter than object C? Circle whether object C is longer than or shorter than object A.

See and Show

Mathematical PRACTICE

You can compare the **lengths** of objects.
The paintbrush is longer than the pencil.
The pencil is longer than the tube of paint.

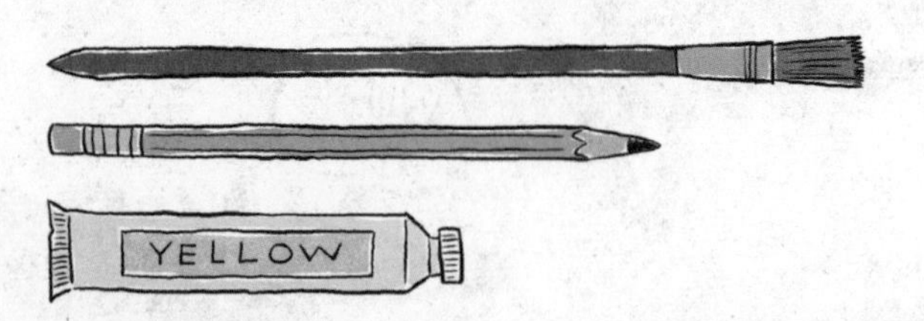

Helpful Hint
The length of objects can be shorter than or longer than each other.

Is the tube of paint **longer** than or **shorter** than the paintbrush?

longer than   shorter than

Compare the objects. Circle the correct answer.

1. The pair of glasses is shorter than the marker.
The marker is shorter than the picture.

Is the picture longer than or shorter than the pair of glasses?

longer than   shorter than

Talk Math How can you tell if an object is longer than or shorter than another object?

Name

On My Own

Compare the objects. Circle the correct answer.

2. The crayon is longer than the paper clip. The scissors are longer than the crayon.

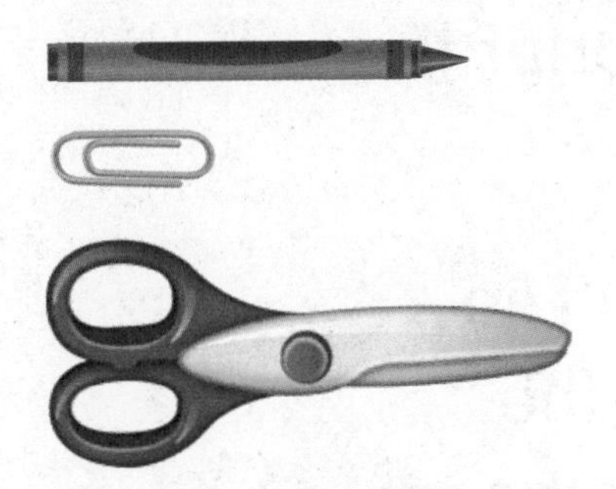

Is the paper clip longer than or shorter than the scissors?

longer than   shorter than

3. The eraser is shorter than the glue stick. The glue stick is shorter than the pencil.

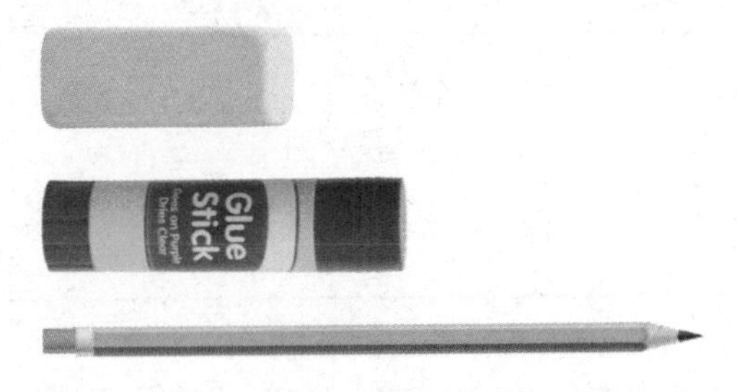

Is the pencil longer than or shorter than the eraser?

longer than   shorter than

Problem Solving

Compare the objects. Circle the correct answer.

4. The chalk is shorter than the pen.
The paper clip is shorter than the chalk.
Is the pen longer than or shorter than
the paper clip?

longer than shorter than

5. The table is longer than the book. The
book is longer than the eraser. Is the
eraser longer than or shorter than the table?

longer than shorter than

Write Math A marker is longer than a glue stick.
A glue stick is longer than a paper
clip. Is the paper clip longer than or
shorter than the marker? Explain.

Name ....

My Homework

Lesson 1
Compare Lengths

Homework Helper

Need help? connectED.mcgraw-hill.com

You can compare the lengths of objects.

The carrot is longer than the pea pod.
The pea pod is longer than the apple.

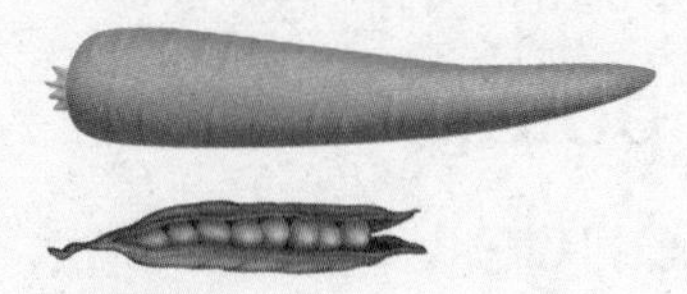

Is the carrot longer than or shorter than the apple?

shorter than

Practice

Compare the objects. Circle the correct answer.

1. The juice pop is shorter than the corn cob.
The cracker is shorter than the juice pop.

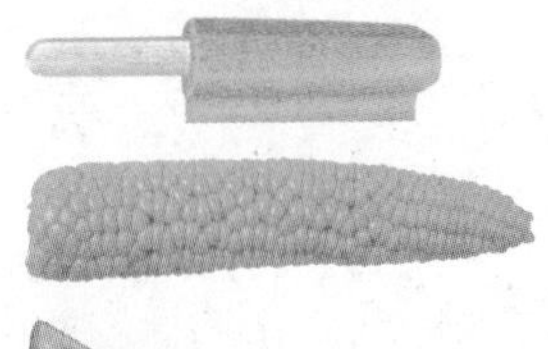

Is the cracker longer than or shorter than the corn cob?

longer than   shorter than

Compare the objects. Circle the correct answer.

2. The piece of candy is shorter than the sandwich. The sandwich is shorter than the celery.

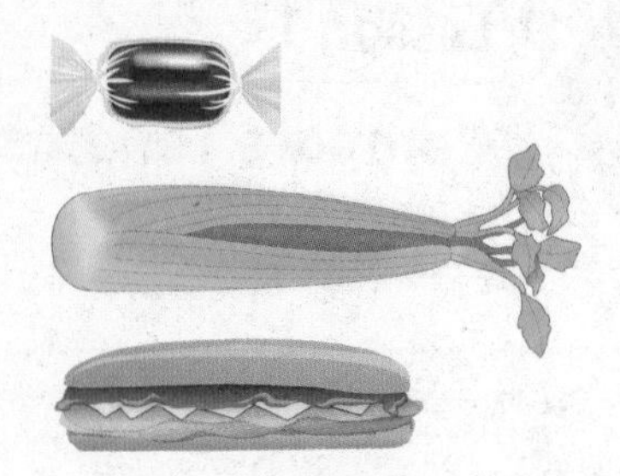

Is the celery longer than or shorter than the piece of candy?

longer than　　　shorter than

3. A grape is shorter than a potato. The potato is shorter than a hot dog. Is a grape longer than or shorter than a hot dog?

longer than　　　shorter than

Vocabulary Check

Draw lines to match.

4. **longer**

5. **shorter**

Math at Home Give your child three objects of various lengths. Ask him or her to compare the lengths of the objects.

Name ....

Compare and Order Lengths

Lesson 2

ESSENTIAL QUESTION
How do I determine length and time?

We can help!

Explore and Explain

Watch

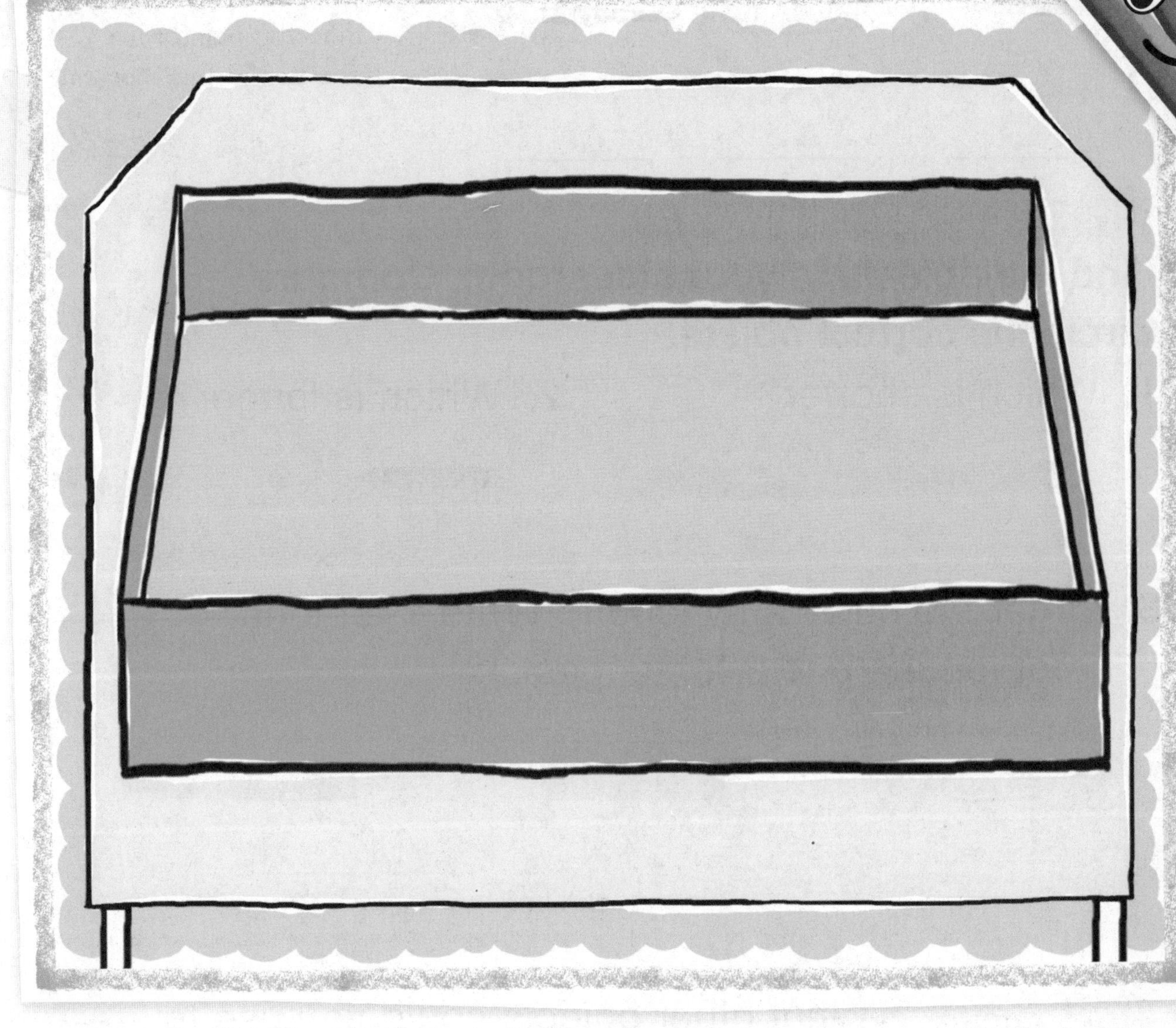

Teacher Directions: Find two objects that will fit in the box. Draw them in the box. Circle the object that is longer. Put an X on the object that is shorter.

See and Show

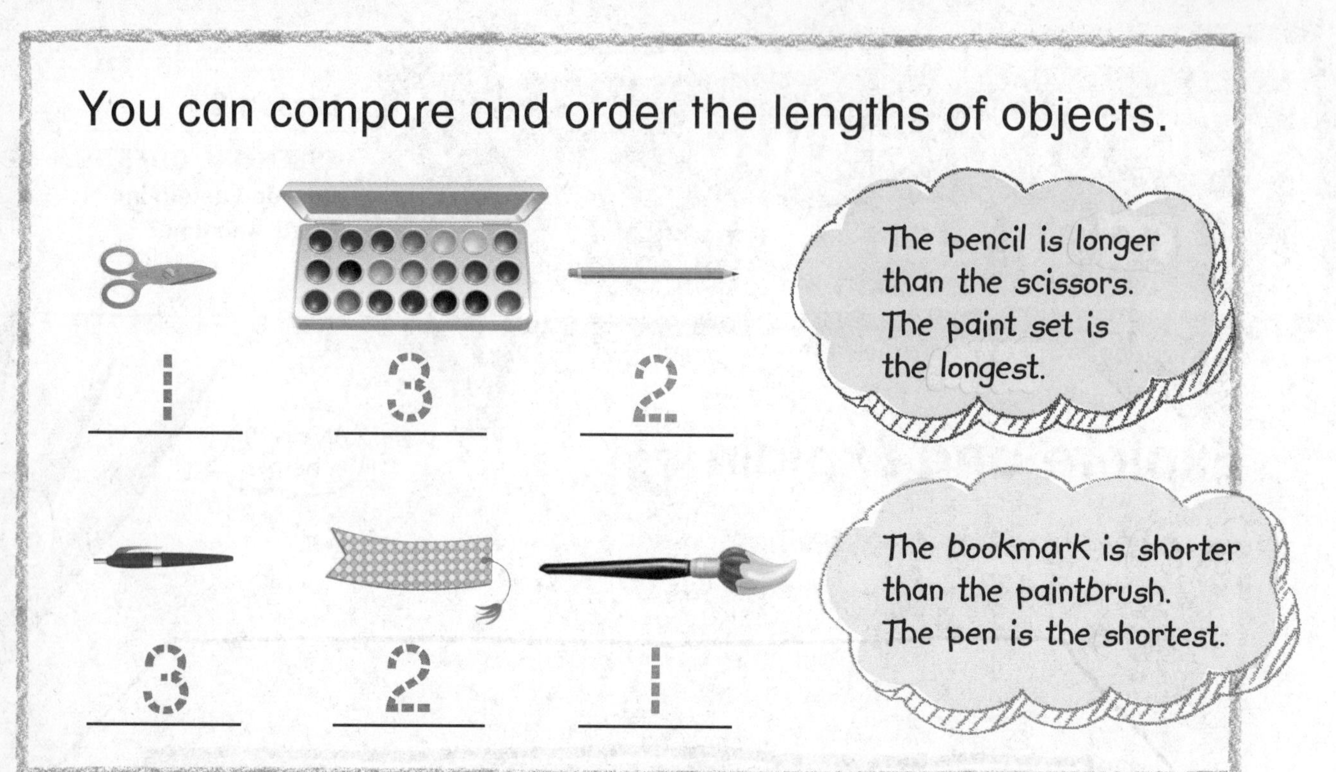

Find the objects in your classroom. Compare. Circle the correct object.

1. Which is shorter?

2. Which is longer?

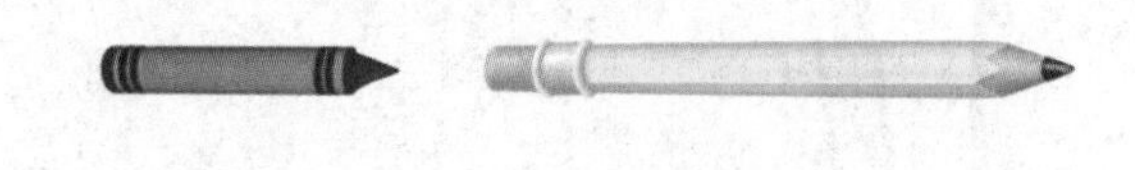

3. Order the objects by length. Write 1 for long, 2 for longer, and 3 for longest.

_______ _______ _______

What other objects could you use to compare lengths?

Name

On My Own

Find the objects in your classroom. Compare. Circle the correct object.

4. Which is shorter?

5. Which is longer?

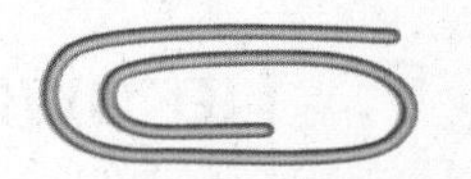

6. Order the objects by length. Write 1 for short, 2 for shorter, and 3 for shortest.

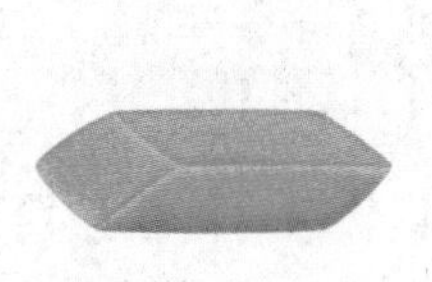

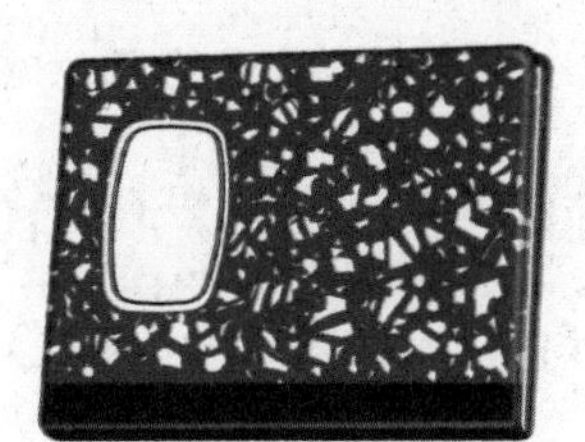

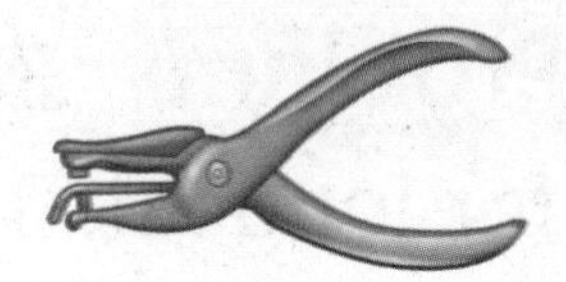

______ ______ ______

7. Order the objects by length. Write 1 for long, 2 for longer, and 3 for longest.

______ ______ ______

Problem Solving

8. John has three objects. How should he order the objects from shortest to longest?

______ ______ ______

9. Tia owns three iguanas. How should she order the iguanas from longest to shortest?

______ ______ ______

HOT Problem Zoe writes 1 for long, 2 for longer, and 3 for longest. Tell why Zoe is wrong. Make it right.

3 2 1

Name ....

My Homework

Lesson 2
Compare and Order Lengths

Homework Helper

Need help? connectED.mcgraw-hill.com

You can compare and order the lengths of objects.

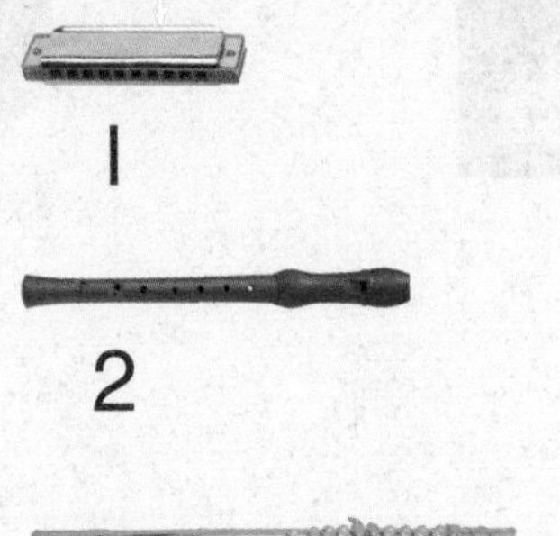

Helpful Hint

The objects are ordered by length. 1 shows the long object, 2 shows the longer object, and 3 shows the longest object.

Practice

Compare. Circle the shorter object.

1.

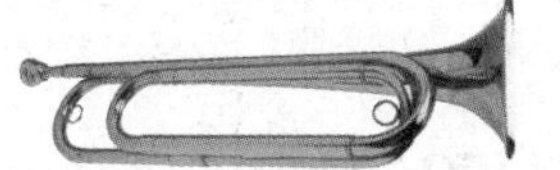

2.

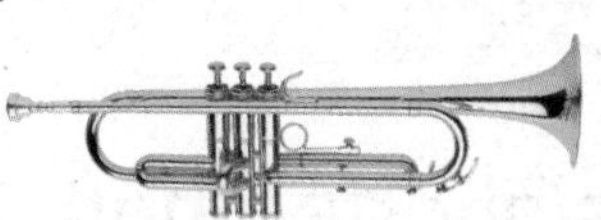

3. Order the objects by length. Write 1 for short, 2 for shorter, and 3 for shortest.

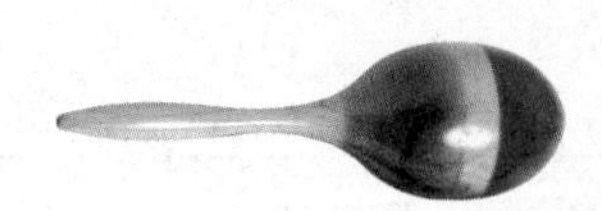

______ ______ ______

Compare. Circle the longer object.

4.

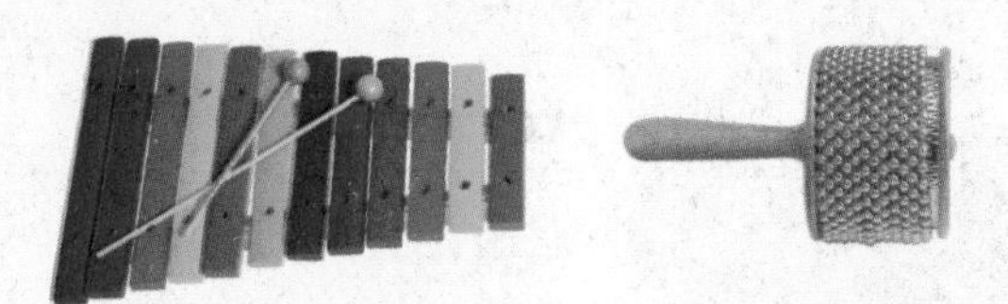

5.

6. Order the objects by length. Write 1 for long, 2 for longer, and 3 for longest.

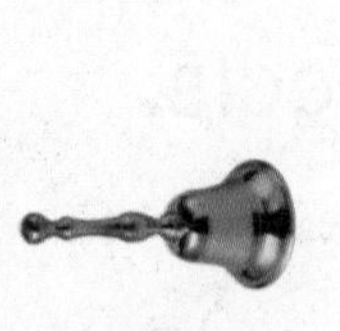

______ ______ ______

7. Kiah has a music book and a recorder. Which object is longer? Circle it.

Test Practice

8. Which is shorter than this instrument?

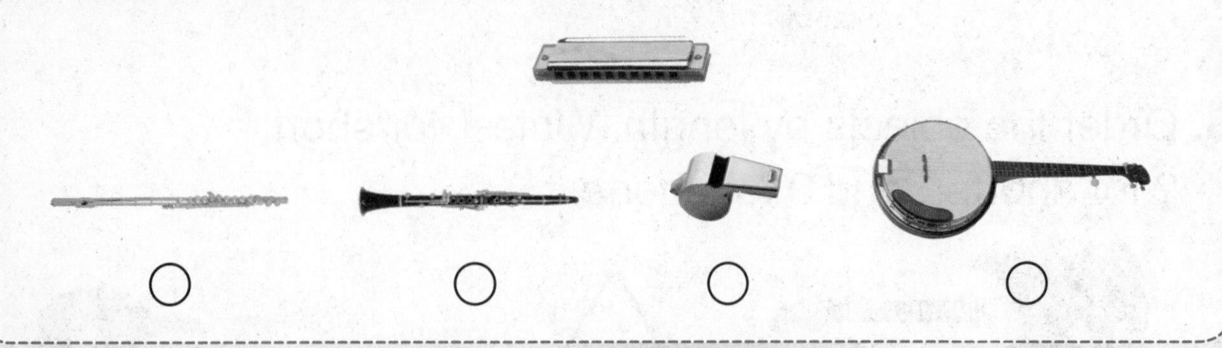

○ ○ ○ ○

Math at Home Find two objects in the kitchen. Have your child describe them by comparing their lengths.

Name

Nonstandard Units of Length

Lesson 3

ESSENTIAL QUESTION
How do I determine length and time?

Explore and Explain

Watch

Tools

Does yellow look good on me?

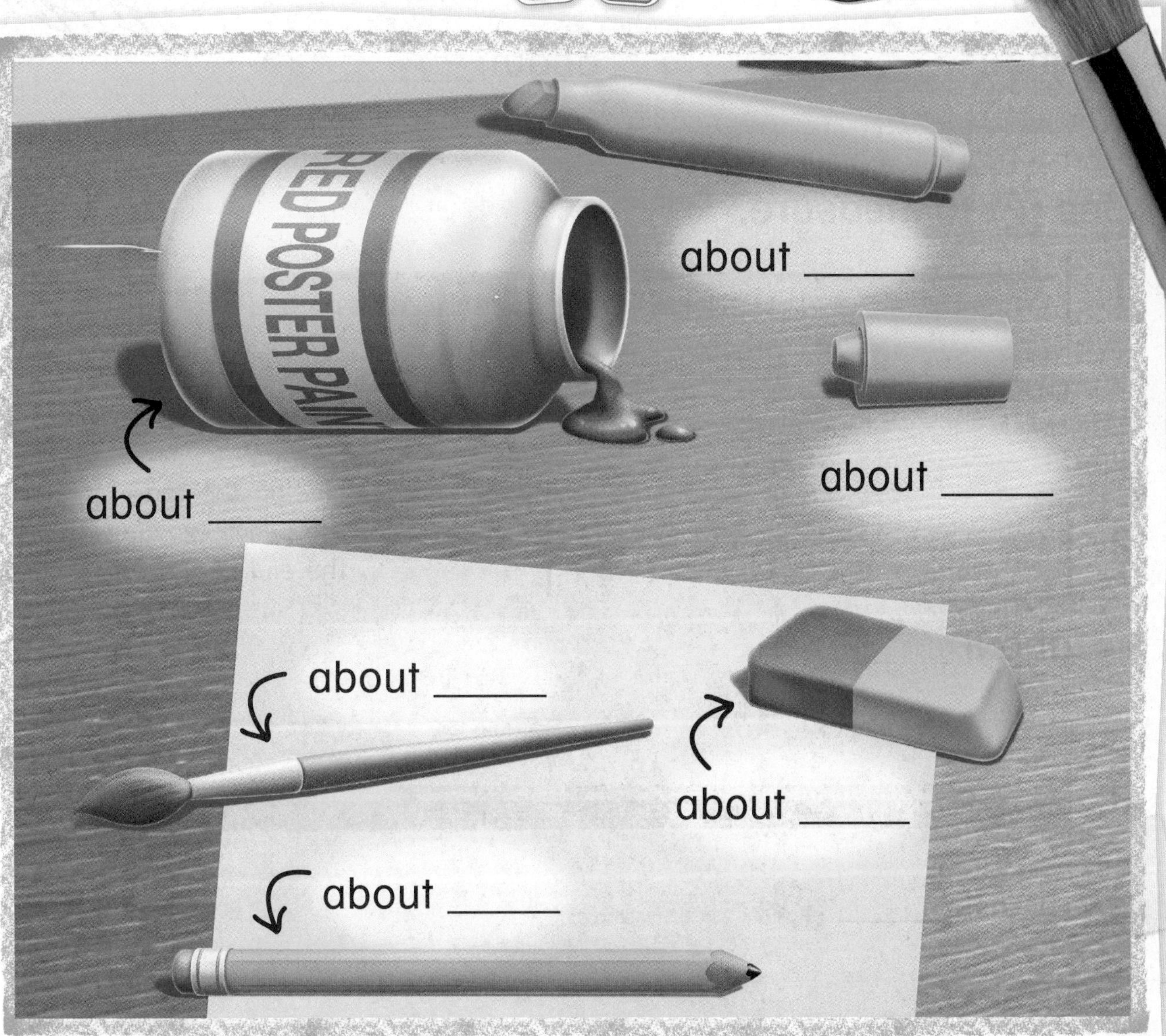

Teacher Directions: Use [cube] to measure. Write the length of each object.

See and Show

Mathematical PRACTICE

You can **measure** to find the length of an object. Each cube or paper clip is one **unit**.

about 3 (cubes)   about 2 (cubes)

Helpful Hint
You can measure objects using cubes, paper clips, or pennies.

Use (cube) to measure.

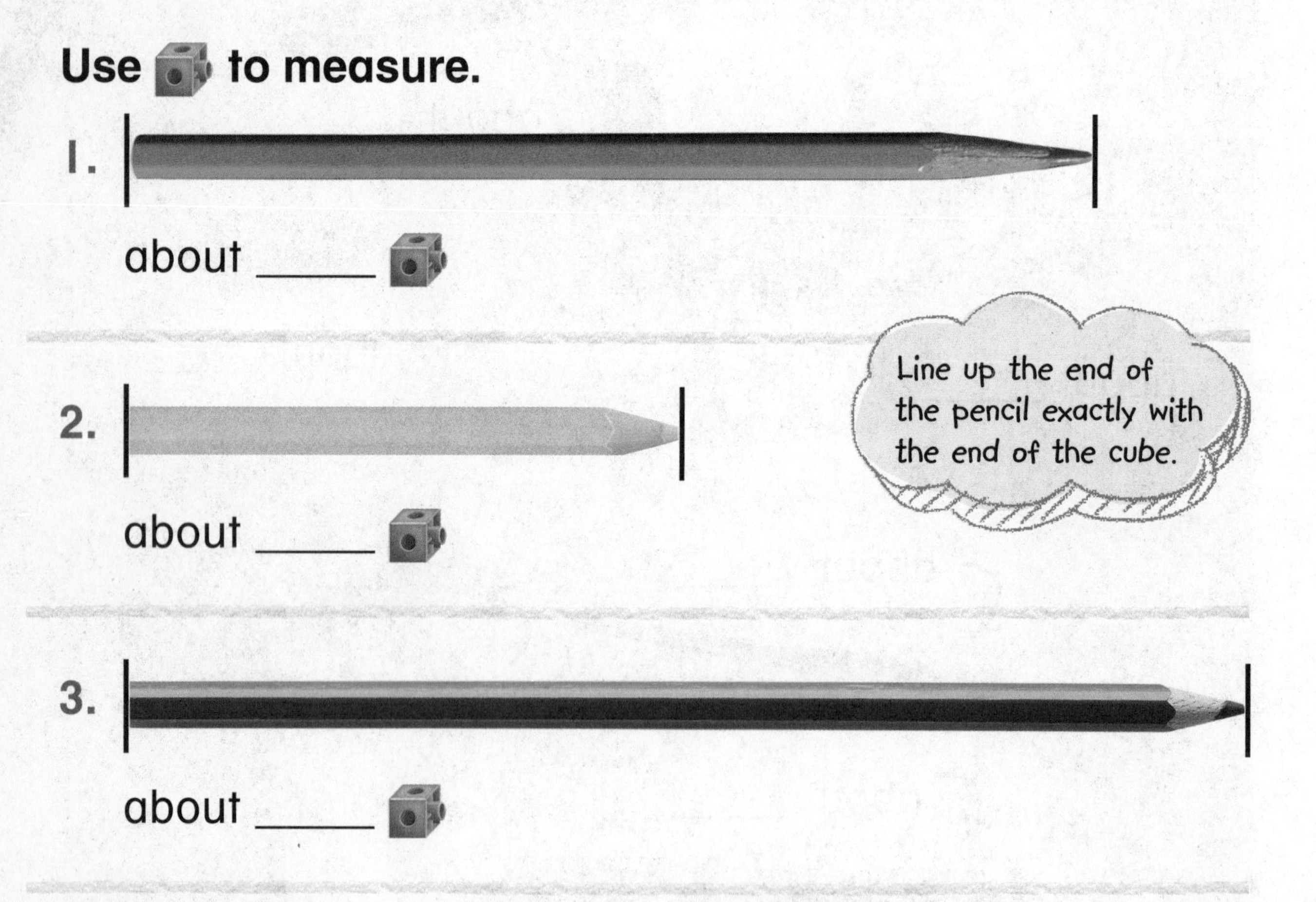

1. about _____ (cubes)

2. about _____ (cubes)

Line up the end of the pencil exactly with the end of the cube.

3. about _____ (cubes)

Talk Math How can you tell which pencil on this page is the longest?

Name ______________________

I can help!

CCSS

On My Own

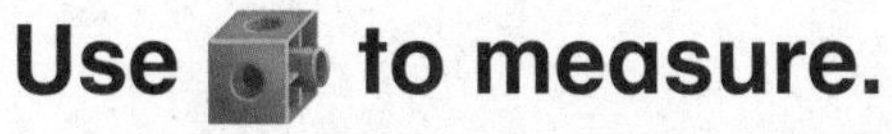

Use [cube] to measure.

4. 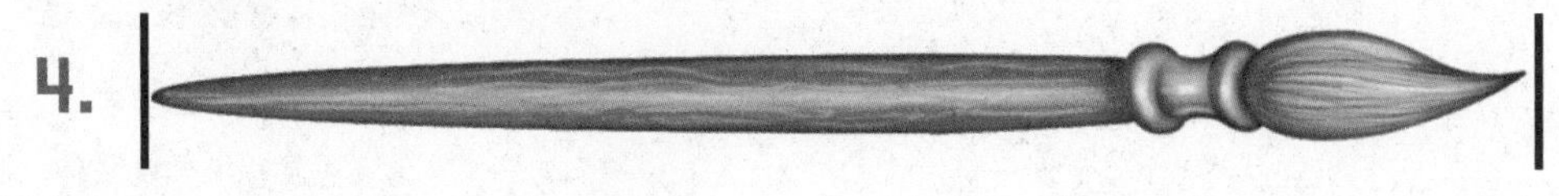

about ______ [cube]

5. 

about ______ [cube]

6. 

about ______ [cube]

7. 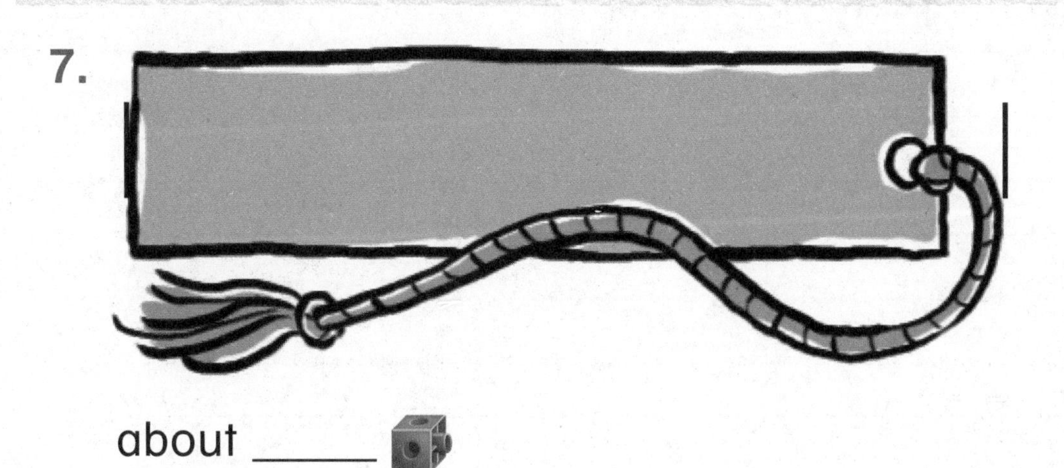

about ______ [cube]

8. Write the name of an object that is about 4 cubes long.

____

Problem Solving

9. About how many cubes long is the pencil? Circle the correct answer.

2 cubes   5 cubes   10 cubes

10. Draw an object that is about 7 cubes long.

HOT Problem A book is 10 cubes long. A pencil box is 3 cubes shorter than the book. How long is the pencil box? Explain.

Name

My Homework

Lesson 3
Nonstandard Units of Length

Homework Helper

Need help? connectED.mcgraw-hill.com

You can measure the length of an object using pennies.

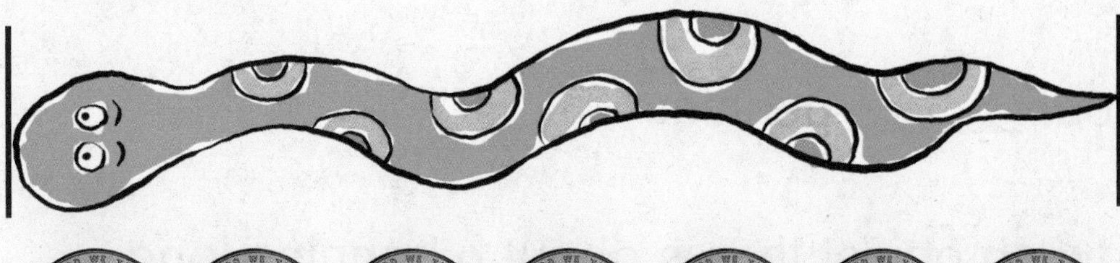

The snake is about 7 pennies long.

Helpful Hint
Each penny is one unit.

Practice

Use pennies to measure.

1. 

about ______ pennies

2. 

about ______ pennies

Use pennies to measure.

3.

about ______ pennies

4.

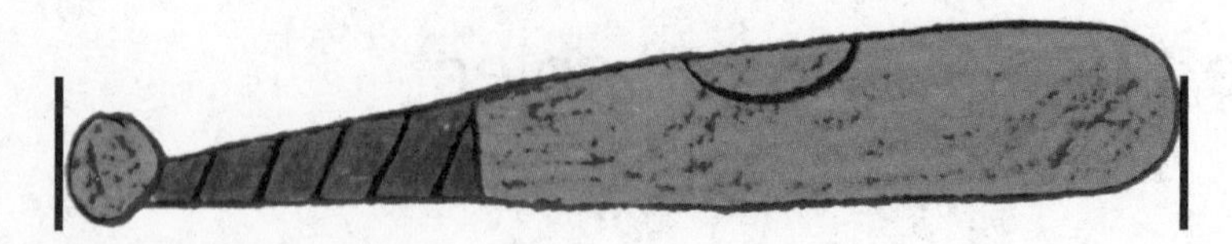

about ______ pennies

5. Draw an object that is about 6 pennies long.

Vocabulary Check

Complete each sentence.

unit **measure**

6. You can ______________ objects using pennies or cubes.
7. Each cube or penny stands for one ______________.

Math at Home Have your child use a nonstandard unit (such as a penny or macaroni) to measure and compare objects.

Name

Problem Solving

STRATEGY: Guess, Check, and Revise

Lesson 4
ESSENTIAL QUESTION
How do I determine length and time?

The carrot is less than 10 cubes long. It is more than 2 cubes long. About how many cubes long is the carrot?

1 Understand Underline what you know. Circle what you need to find.

2 Plan How will I solve the problem?

3 Solve I will guess, check, and revise.

Guess: about 4 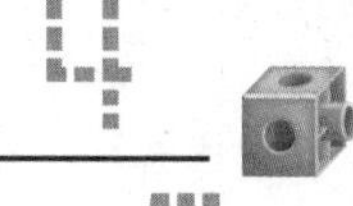

Measure: about 5 

4 Check Is my answer reasonable? Explain.

Practice the Strategy

The container is less than 9 cubes long.
It is more than 1 cube long.
About how long is the container?

1 Understand Underline what you know.
Circle what you need to find.

2 Plan How will I solve the problem?

3 Solve I will . . .

Guess: about _____

Measure: about _____

4 Check Is my answer reasonable? Explain.

Name ....

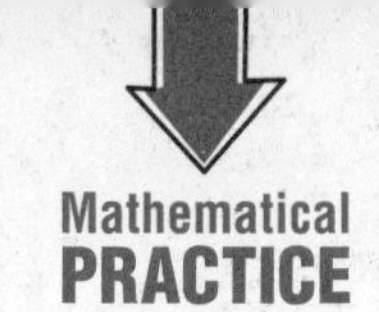

CCSS

Apply the Strategy

About how many (cube) long is the object?
Guess. Then measure. Revise if needed.

1. 

Guess: about _____ (cube)

Measure: about _____ (cube)

2. 

Guess: about _____ (cube)

Measure: about _____ (cube)

3. 

Guess: about _____ (cube)

Measure: about _____ (cube)

Review the Strategies

Choose a strategy

- Guess, check, and revise.
- Draw a diagram.
- Act it out.

4. The pea pod is more than 1 cube long. It is less than 4 cubes long. About how many cubes long is the pea pod?

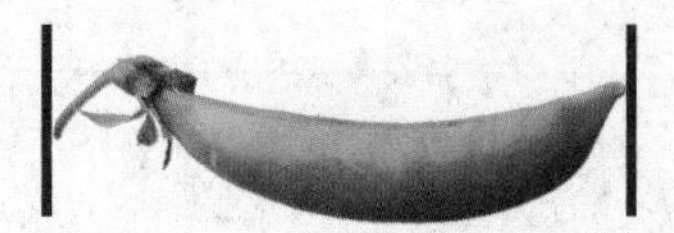

about _____

5. A watermelon grows 1 cube longer each day. On Monday, it was 6 cubes long. How many cubes long will the watermelon be on Wednesday?

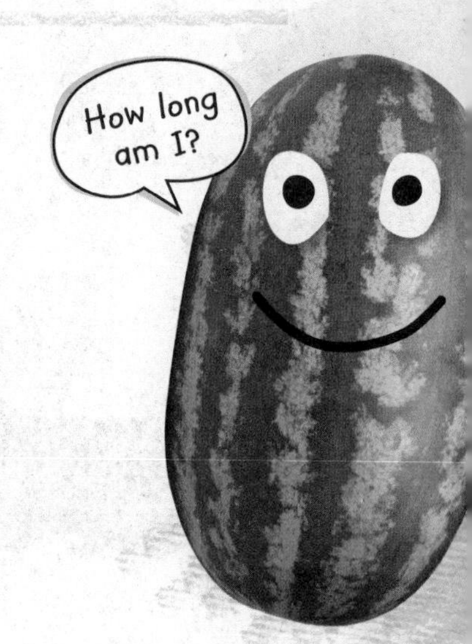

Days	Cubes
Monday	6
Tuesday	
Wednesday	

_____

6. Sergio picks a bean that is 5 cubes long. Perla picks a bean that is 2 cubes long. How much longer is the bean that Sergio picked?

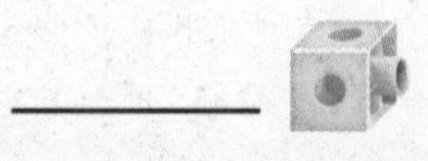

Name

My Homework

Lesson 4
Problem Solving: Guess, Check, and Revise

Homework Helper

Need help? connectED.mcgraw-hill.com

The baseball bat is less than 6 pennies long. It is more than 1 penny long. About how many pennies long is the bat?

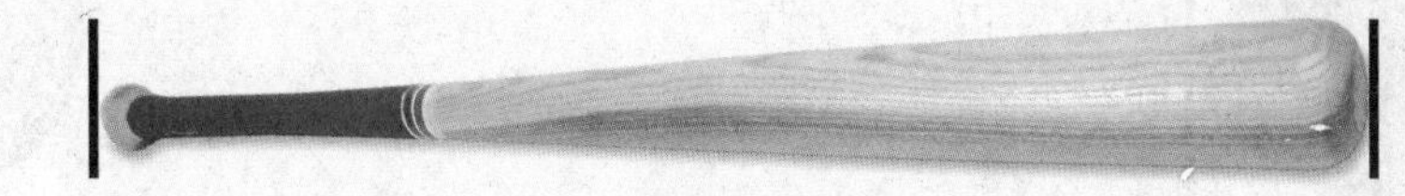

1 Understand Underline what you know.
Circle what you need to find.

2 Plan How will I solve the problem?

3 Solve I will guess, check, and revise.

Guess: about 4 pennies

Measure: about 5 pennies

4 Check Is my answer reasonable?

Problem Solving

About how many pennies long is the object? Guess. Then measure. Revise if needed.

1.

Guess: about _____ pennies

Measure: about _____ pennies

2. 

Guess: about _____ pennies

Measure: about _____ pennies

3. 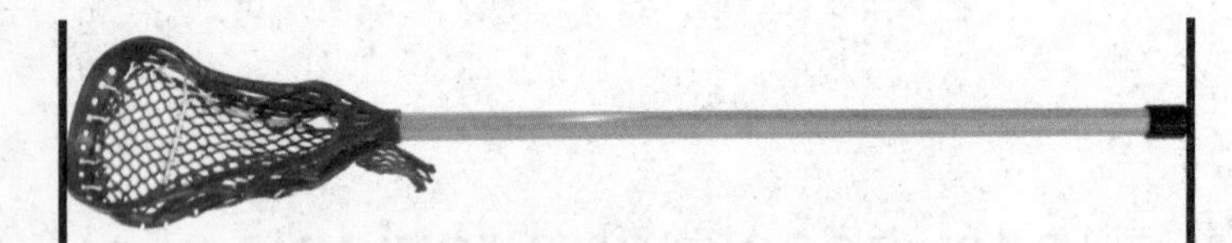

Guess: about _____ pennies

Measure: about _____ pennies

Math at Home Ask your child to guess the length of his or her own shoe in pennies. Then check using actual pennies.

Name

Check My Progress

Vocabulary Check 

Complete each sentence.

length **unit**

1. You can use cubes and paper clips to measure. Each cube or paper clip is one ________.

2. You can measure how long an object is, or its ________.

Concept Check

Compare. Circle the shorter object.

3.

4. 

5. Order the objects by length. Write 1 for short, 2 for shorter, and 3 for shortest.

____ ____ ____

6. Order the objects by length. Write 1 for long, 2 for longer, and 3 for longest.

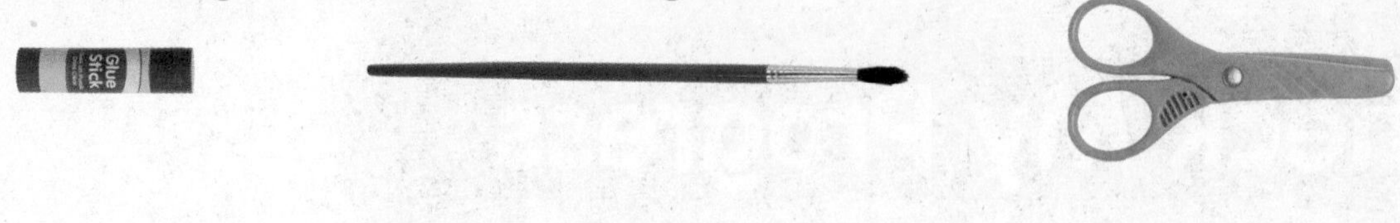

______ ______ ______

Measure using [cube]. Write how many.

7.

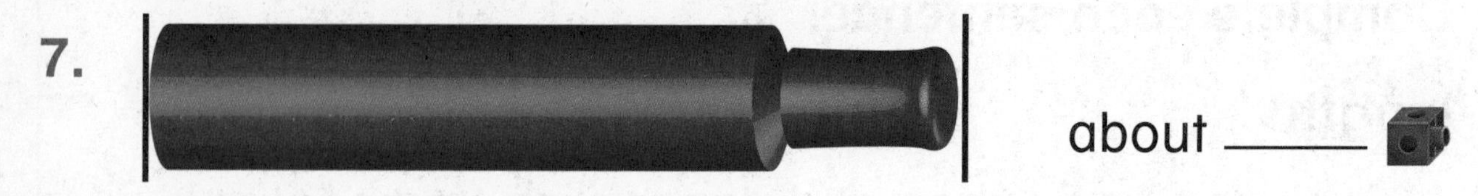

about ______ [cube]

8. Gia has a box, a globe, and an apple. The box is longer than the globe. The globe is longer than the apple. Is the apple longer than or shorter than the box?

longer than shorter than

Test Practice

9. What object is longer than the pencil?

○ ○ ○ ○

Name

Time to the Hour: Analog

Lesson 5

ESSENTIAL QUESTION
How do I determine length and time?

Explore and Explain

Teacher Directions: Use [clock] to model the times given. Robbie has band practice that starts at 3 o'clock. Show that time on the clock. The practice is over at 5 o'clock. Show that time on the clock. Trace the hands to show the time on this clock.

See and Show

Mathematical PRACTICE

There are two kinds of clocks.
This is an **analog clock**.

The **hour hand** is shorter.
It points to the **hour**.

The **minute hand** is longer.
It points to the **minute**.

3 o'clock

The hour hand is on 3. It is 3 o'clock.

Use [clock] to show the time. Tell what time is shown. Write the time.

1. 

_____ o'clock

2. 

_____ o'clock

3. 

_____ o'clock

4. 

_____ o'clock

Talk Math Where are the hour hand and the minute hand when it is 4 o'clock?

Name ______________________________

Time to practice!

CCSS

On My Own

Use [clock] to show the time.
Tell what time is shown. Write the time.

5.

______ o'clock

6.

______ o'clock

7.

______ o'clock

8.

______ o'clock

9.

______ o'clock

10.

______ o'clock

11.

______ o'clock

12.

______ o'clock

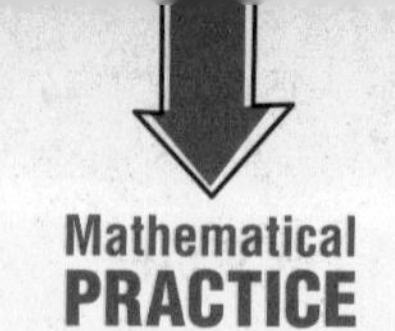

Problem Solving

Solve. Write the time. Draw the hands on the clock. Use 🕘 to help.

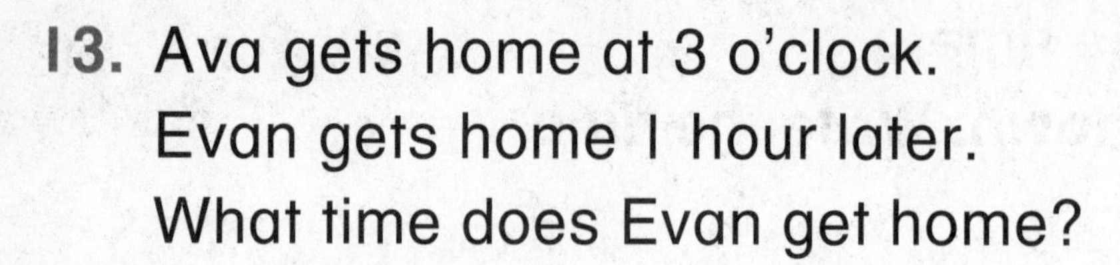

13. Ava gets home at 3 o'clock.
Evan gets home 1 hour later.
What time does Evan get home?

______ o'clock

14. Roger begins reading books at 7 o'clock. He reads for one hour. What time does he stop?

______ o'clock

HOT Problem Antonio tried to set the hands on his clock for 9 o'clock. Tell why Antonio is wrong. Make it right.

____

____

____

Name ....

My Homework

Lesson 5
Time to the Hour: Analog

Homework Helper

Need help? connectED.mcgraw-hill.com

On an analog clock, the hour hand is shorter.
The minute hand is longer.

11 o'clock

6 o'clock

Practice

Tell what time is shown. Write the time.

1.

_____ o'clock

2.

_____ o'clock

3.

_____ o'clock

4.

_____ o'clock

Tell what time is shown. Write the time.

5. 

_____ o'clock

6. 

_____ o'clock

7. Mr. Smith's class starts at 9 o'clock. It ends one hour later. What time is it over?

_____ o'clock

8. Miguel has soccer practice at 7 o'clock. It lasts for one hour. What time is the practice over?

_____ o'clock

Vocabulary Check 

Complete the sentences.

minute hand **hour hand**

9. On an analog clock, the _______________ is shorter.

The _______________ is longer.

Math at Home Ask your child to say the times to the hour in order, beginning with 1 o'clock (1 o'clock, 2 o'clock, 3 o'clock, and so on).

Name ....

Time to the Hour: Digital

Lesson 6

ESSENTIAL QUESTION
How do I determine length and time?

Explore and Explain

Rise and shine!

Teacher Directions: Emilia wakes up for school at 7 o'clock. Use  to show the time. Write that time on the digital clock.

See and Show

Another type of clock is a **digital clock**. A digital clock uses numbers to show the hour and minutes.

The clock shows 2 o'clock.

Use [clock] to show the time. Tell what time is shown. Write the time on the digital clock.

1\.

2\.

3\.

4\.

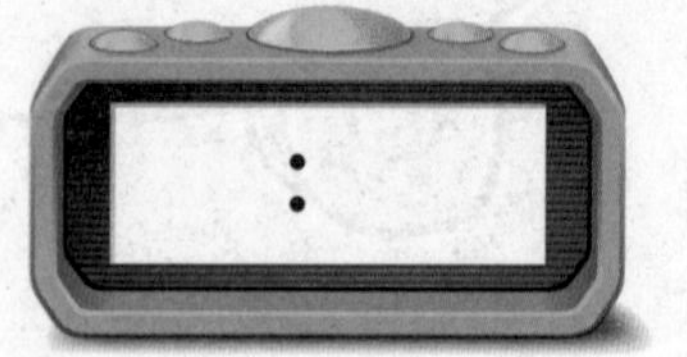

Talk Math How is reading an analog clock the same as reading a digital clock?

Name ....

On My Own

Use to show the time. Tell what time is shown. Write the time on the digital clock.

5.

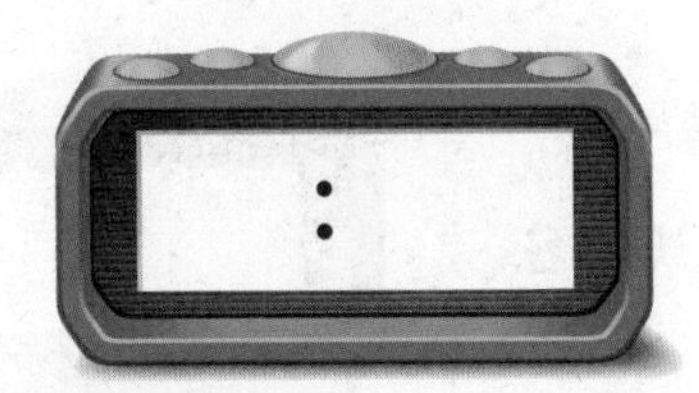

6.

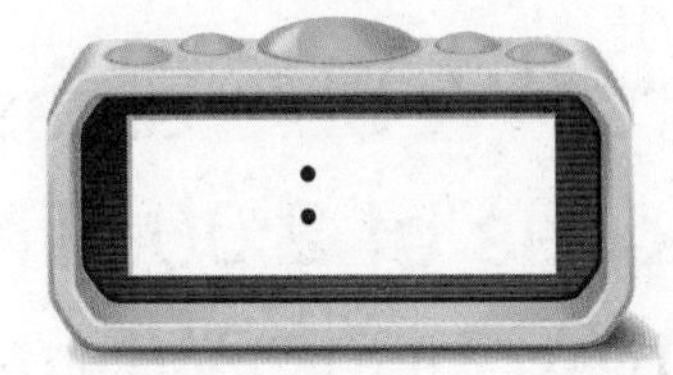

7.

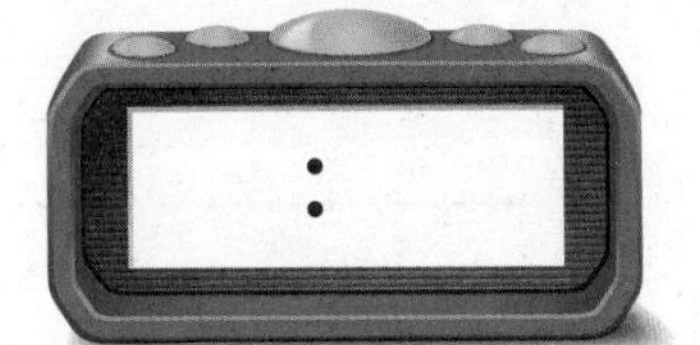

8.

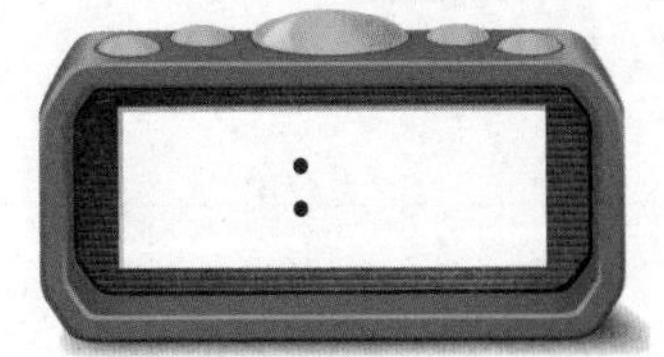

9.

10.

Problem Solving

Solve. Write the time on the digital clock.

11. Rachel goes to the cafeteria to eat at 11:00. She is there for 1 hour. What time does she leave the cafeteria?

_____ o'clock

12. Mrs. Webb's class came in from recess at 2:00. They were at recess for an hour. What time did they go out to recess?

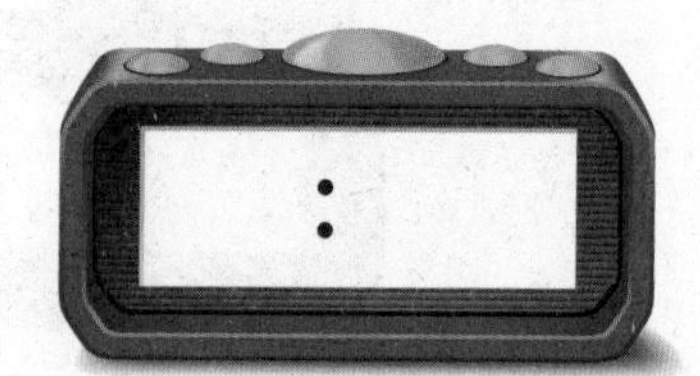

_____ o'clock

Write Math Explain how a digital clock shows time.

Name ....

My Homework

Lesson 6
Time to the Hour: Digital

Homework Helper

Need help? connectED.mcgraw-hill.com

An analog clock uses an hour hand and a minute hand to show time. A digital clock uses numbers to show the hour and minutes.

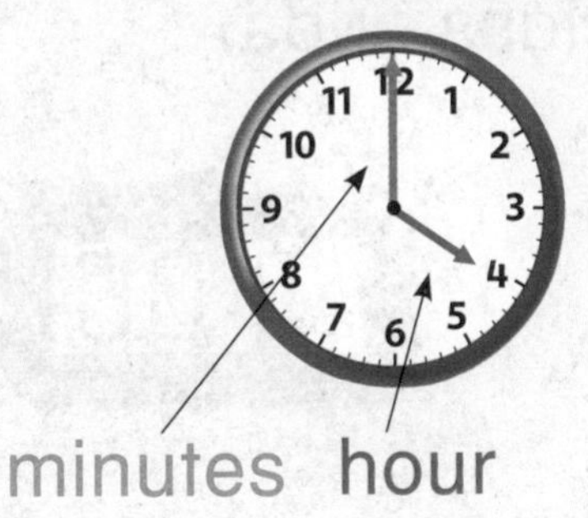

Practice

Tell what time is shown.
Write the time on the digital clock.

1.

2.

Tell what time is shown.
Write the time on the digital clock.

3.

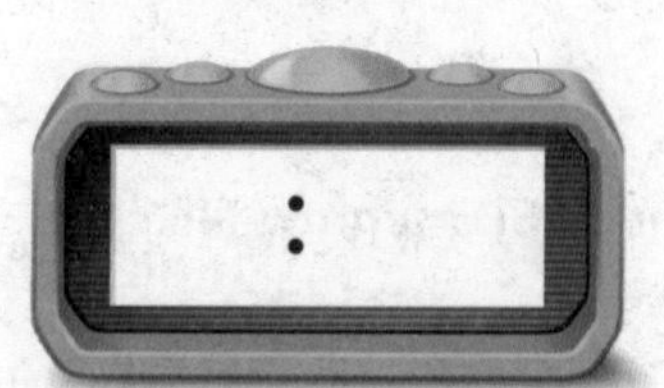

4.

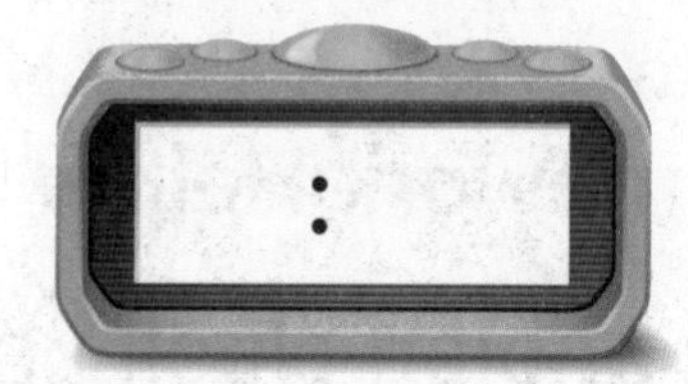

5. Layla's dance class starts at 6 o'clock. It ends 1 hour later. Circle the clock that shows the time dance class ends.

Vocabulary Check

Circle the correct answer.

6. digital clock

Math at Home Set a digital clock to various hours and ask your child to name the hours the clock shows.

Name

Time to the Half Hour: Analog

Lesson 7

ESSENTIAL QUESTION
How do I determine length and time?

Explore and Explain

Teacher Directions: Trace the hour hand red. Trace the minute hand blue. Use  to show the time. Write the time shown on the clock.

See and Show

Mathematical
PRACTICE

An analog clock can show time to the **half hour**.
A half hour is 30 minutes.

Helpful Hint
The hour hand points between the 4 and 5. The minute hand points to the 6. It is 4:30.

It is half past 4 or 4:30.

Use . Tell what time is shown. Write the time.

1.

half past ______

2.

half past ______

3.

half past ______

4.

half past ______

Talk Math It is half past 8. Explain what half past means.

Name ....

CCSS

On My Own

Use . Tell what time is shown. Write the time.

5. 

half past _____

6. 

half past _____

7. 

half past _____

8. 

half past _____

9. 

half past _____

10. 

half past _____

11. 

half past _____

12. 

half past _____

Problem Solving

Mathematical PRACTICE

13. Jen's class is leaving for a field trip to the zoo at 10 o'clock. The class arrives at the zoo 30 minutes later. What time do they arrive at the zoo?

half past ______

14. Nate will go to football practice at half past 4. Show this time on the clock. If practice lasts one hour, what time will it end? Write the time.

half past ______

Write Math What is one difference between the minute hand and hour hand?

____

____

____

Name ....

My Homework

Lesson 7
Time to the Half Hour: Analog

Homework Helper

Need help? connectED.mcgraw-hill.com

An analog clock can show time to the half hour.

half past 12

half past 7

Helpful Hint
One half hour is 30 minutes. It is also called half past.

Practice

Tell what time is shown. Write the time.

1. 

half past _____

2. 

half past _____

3. 

half past _____

4. 

half past _____

Tell what time is shown. Write the time.

5. 

half past _____

6. 

half past _____

7. Tim has swimming practice at 7 o'clock. The swimming practice ends 30 minutes later. What time does it end?

half past _____

8. Trish's bus picks her up at 9 o'clock. She arrives at school 30 minutes later. What time does she arrive at school?

half past _____

Vocabulary Check

Circle the matching time.

9. A **half hour** or half past 12 o'clock.

Math at Home Give your child a time to the hour. Have him or her tell you where the clock hands would be for half past that hour.

Name ....

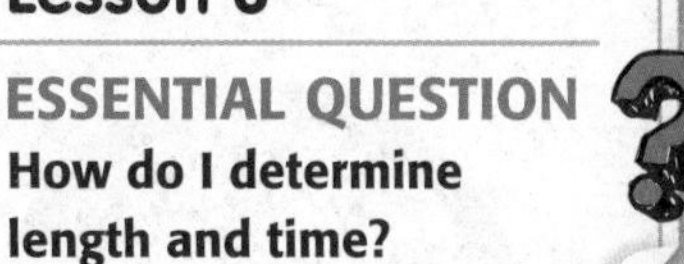

Time to the Half Hour: Digital

Lesson 8

ESSENTIAL QUESTION
How do I determine length and time?

Ready? Set? Practice!

Explore and Explain

Teacher Directions: Tyler's football practice ends at half past 4. Use [clock] to show the time. Write that time on the digital clock shown on the phone. Tell a classmate what time is shown on the phone.

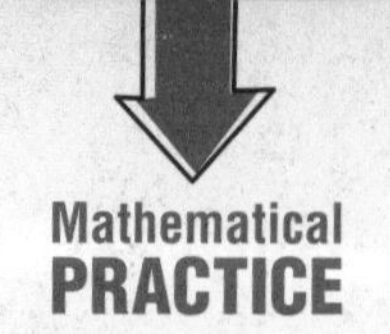

See and Show

A digital clock can also show time to the half hour.

Both of the clocks show half past 6 or 6:30.

Use [clock] to show the time. Tell what time is shown. Write the time on the digital clock.

1.

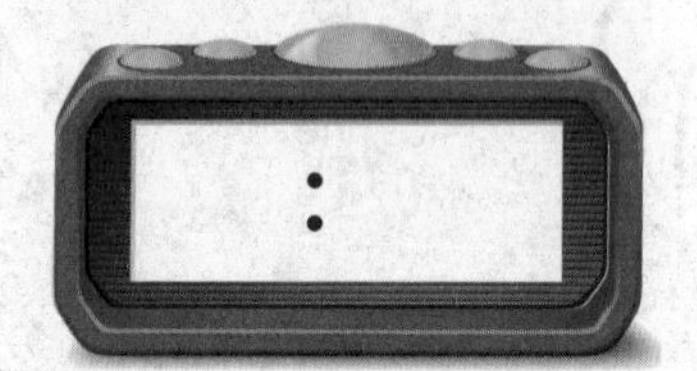

2.

3.

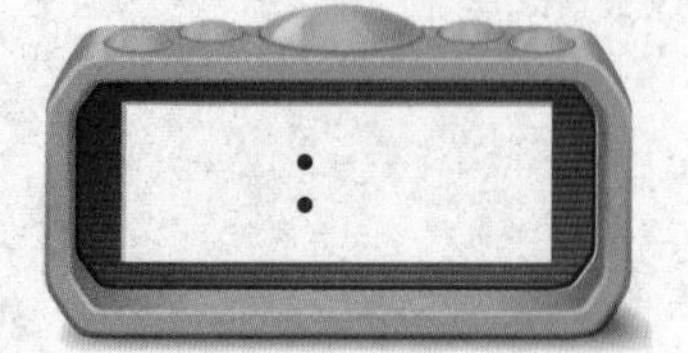

4.

Talk Math How is half past 10 shown on a digital clock?

Name

It's up to you!

CCSS

On My Own

Use [clock] to show the time. Tell what time is shown. Write the time on the digital clock.

5.

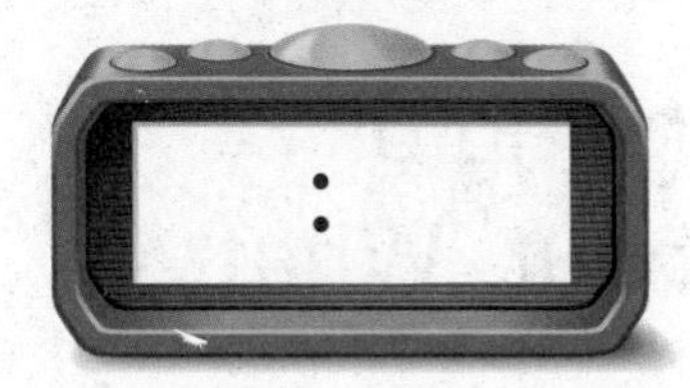

6.

7.

8.

9.

10.

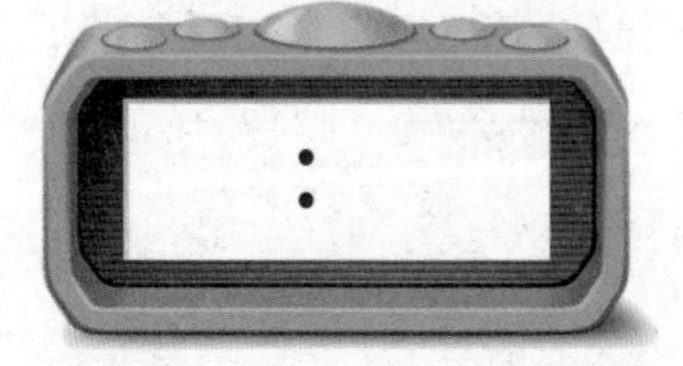

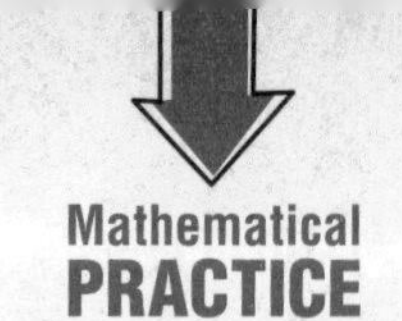

Problem Solving

Mathematical PRACTICE

Solve. Write the time on the digital clock.

11. Mrs. Johnson's class has art at 9:30. It lasts 1 hour. What time does art class end?

12. Aidan's school choir will start singing at half past 2. They will sing for 1 hour. What time will they stop singing?

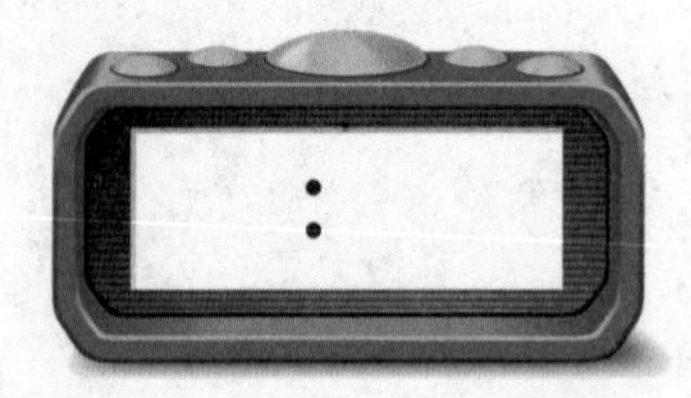

HOT Problem Chase tells his friend that the time on the clock is 12 o'clock. Tell why Chase is wrong. Make it right.

Name ....

My Homework

Lesson 8
Time to the Half Hour: Digital

Homework Helper

Need help? connectED.mcgraw-hill.com

A digital clock can also show time to the half hour.

analog clock **digital clock**

These clocks show half past 7 or 7:30.

Practice

Tell what time is shown.
Write the time on the digital clock.

1.

2.

Tell what time is shown. Write the time on the digital clock.

3.

4.

Solve. Write the time on the digital clock.

5. Will begins his homework at 6 o'clock. He finishes it in a half hour. Write that time on the clock.

Test Practice

6. Which clock shows half past 12?

Math at Home Show your child an analog clock that shows 7:30. Ask him or her to draw a picture of a digital clock that shows the same time.

Name

Time to the Hour and Half Hour

Lesson 9

ESSENTIAL QUESTION
How do I determine length and time?

Explore and Explain

Teacher Directions: A class went to the library at 1:30. Use [clock] to show the time. Find the analog and digital clocks in the picture. Show 1:30 on the analog clock. Write 1:30 on the digital clock.

See and Show

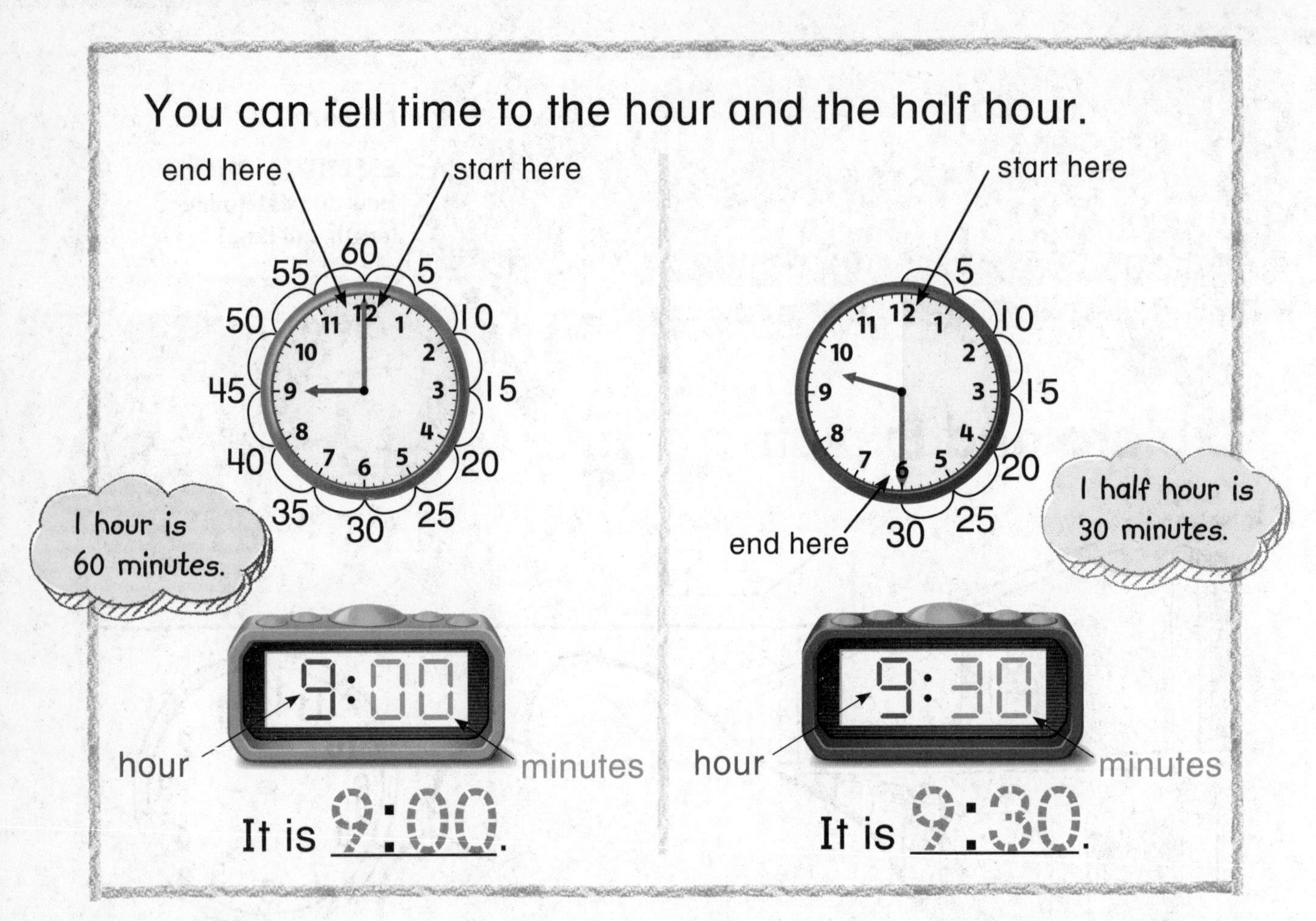

Use [clock] to help. Draw the minute hand to show the time. Write the time on the digital clock.

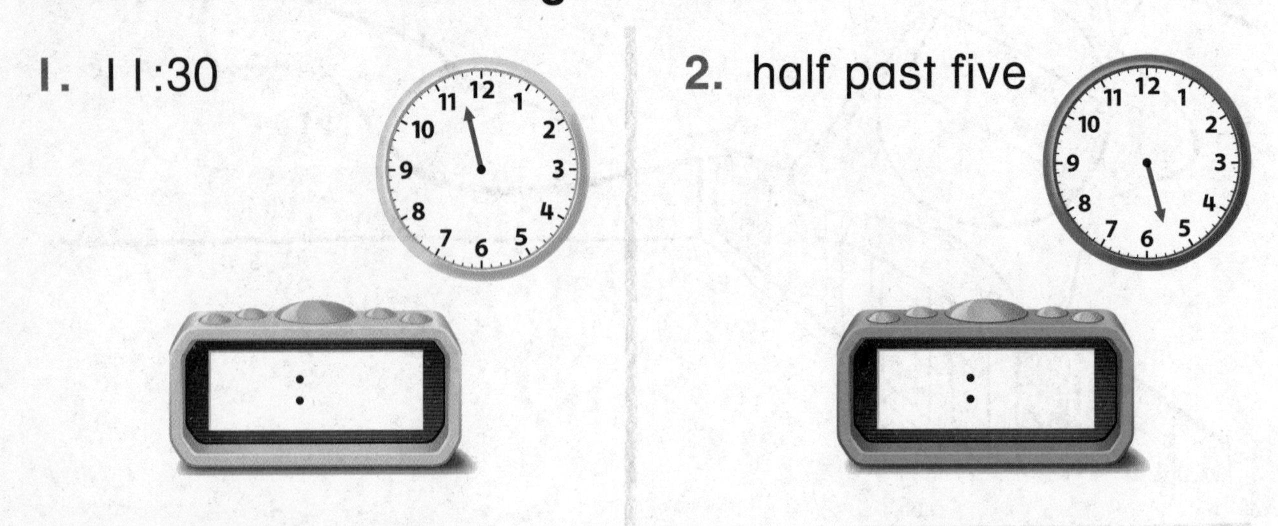

Talk Math What is the difference between an analog clock and a digital clock?

Name

On My Own

Use [clock] to help. Draw the minute hand to show the time. Write the time on the digital clock.

3. 3:30

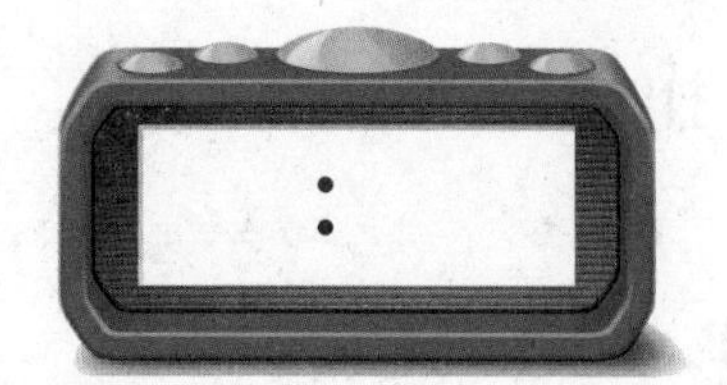

4. nine o'clock

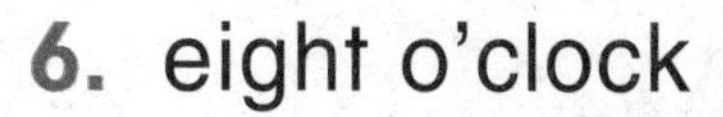

5. half past 1

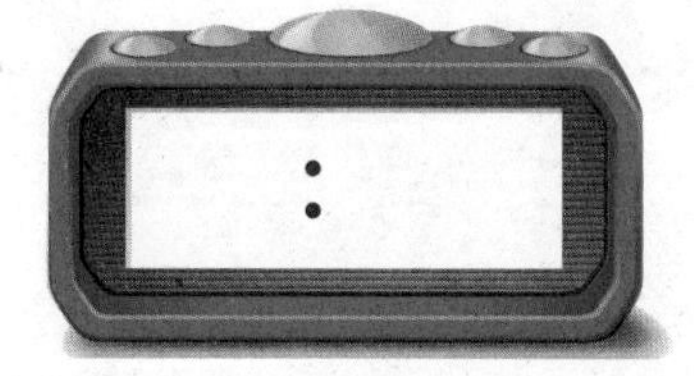

6. eight o'clock

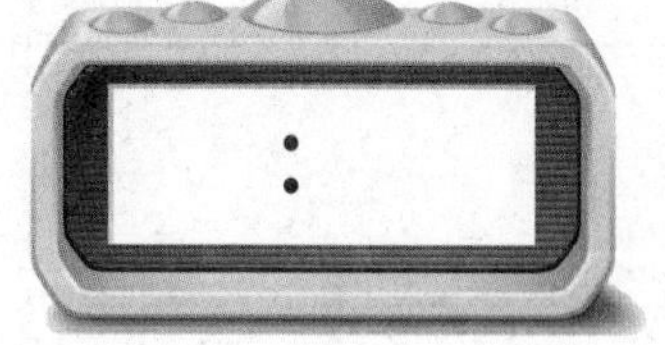

7. Write half past two on the clock.

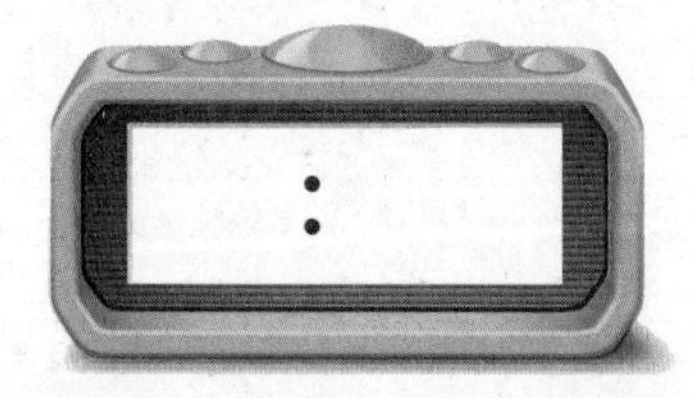

8. Write six o'clock on the clock.

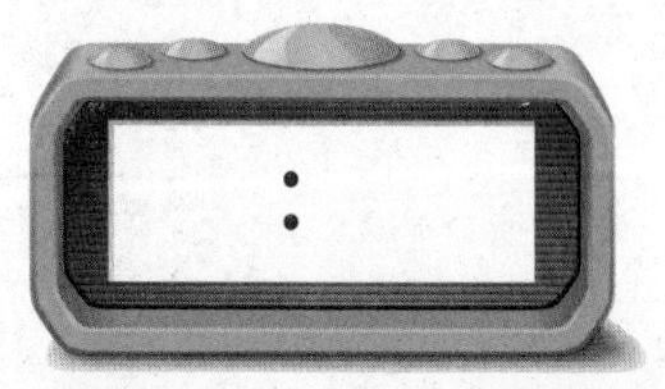

Problem Solving

Mathematical PRACTICE

Solve. Write the time on the clock.

9. The hour hand is between 1 and 2. The minute hand is on 6. What time is it? Draw the hands on the clock.

10. The time on the clock is one hour after 2:00. What time is it? Write the time.

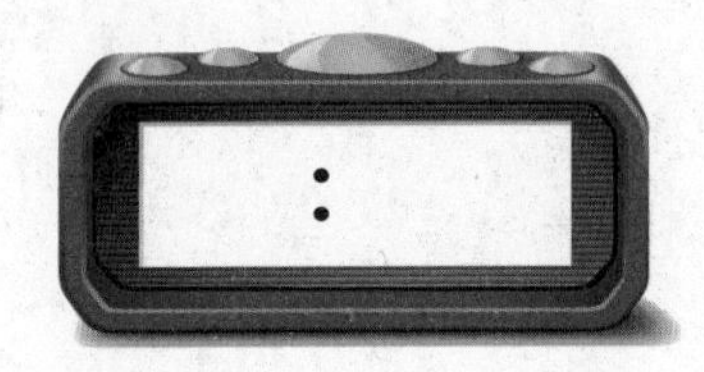

Write Math How many minutes after the hour does this clock show? Explain.

Name

My Homework

Lesson 9
Time to the Hour and Half Hour

Homework Helper

Need help? connectED.mcgraw-hill.com

You can tell time to the hour and the half hour.

It is nine o'clock or 9:00.

It is six-thirty or 6:30.

Practice

Draw the minute hand to show the time.
Write the time on the digital clock.

1. 2:30

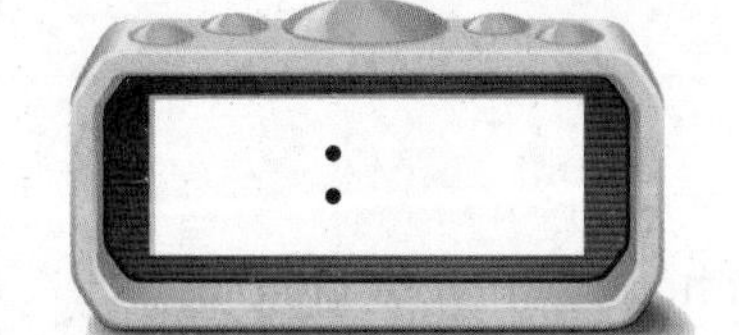

2. half past five

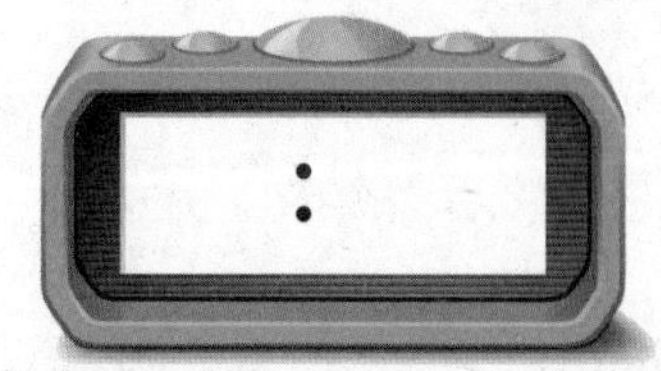

Draw the minute hand to show the time.
Write the time on the digital clock.

3. half past 12

4. 4 o'clock

5. It is half past the hour. My hour hand is between 5 and 6. What time is it? Draw the hands on the clock.

Test Practice

6. What clock shows 8 o'clock?

Math at Home Practice telling time to the hour and half hour with your child by writing the time of familiar events such as mealtime, school time, and bedtime.

Name ______________________

My Review

Chapter 8
Measurement and Time

Vocabulary Check

Complete each sentence.

analog clock **digital clock** **half hour**
length **measure** **minute** **o'clock**

1. A ______________________ is a type of clock that only uses numbers to show time.

2. 30 minutes past the hour is also called a ________________.

3. To find the length of an object, you can ________________ it.

4. You can measure how long an object is, or its ________________.

5. An ______________________ is a clock that uses an hour and minute hand.

6. ________________ is a word used to tell time.

7. There are 60 seconds in 1 ________________.

Concept Check

8. Order the objects by length. Write 1 for long, 2 for longer, and 3 for longest.

_____ _____ _____

About how many cubes long is each object?

9. about _____ cubes

10. about _____ cubes

Write the time on the digital clock.

11. 8 o'clock

12. half past 3

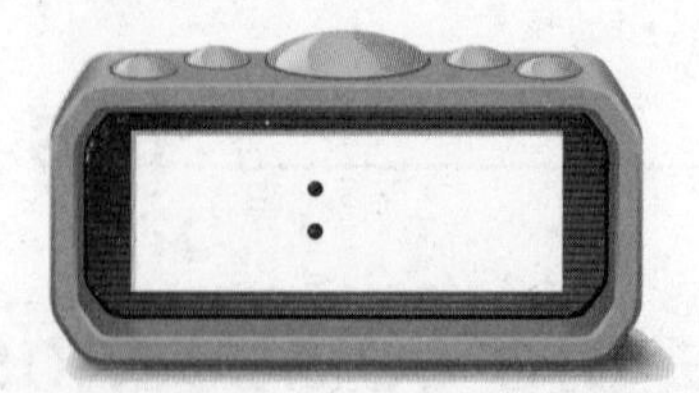

CCSS

Name ....

Problem Solving

Solve. Write the time on the clock.

13. Aaron has swimming practice at 7:00. It lasts 1 hour. What time does swimming practice end?

14. The party starts at 4 o'clock. The party is over 1 hour later. What time does the party end?

Test Practice

15. The hour hand is between 9 and 10. The minute hand is at 6. Which clock is it?

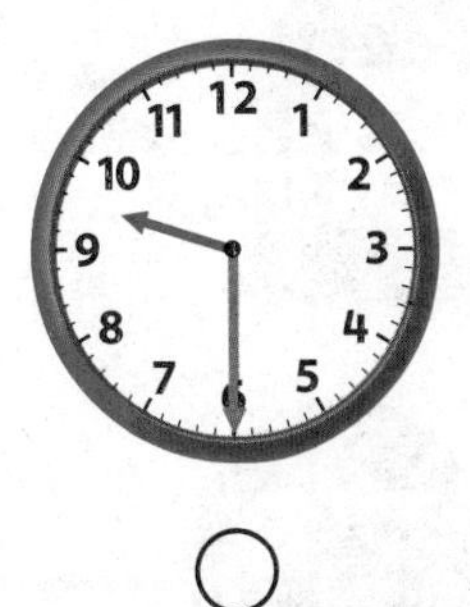 ○

 ○

 ○

 ○

Reflect

Chapter 8

Answering the Essential Question

Show the ways to answer.

Circle the object that is longer. Explain to a classmate how you found the answer.

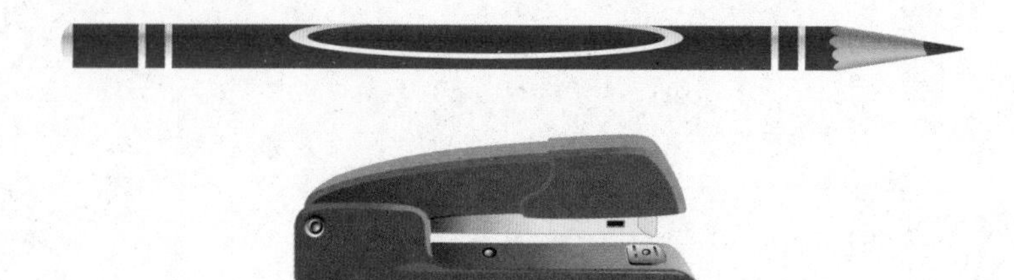

ESSENTIAL QUESTION

How do I determine length and time?

Show the same time on both of the clocks.

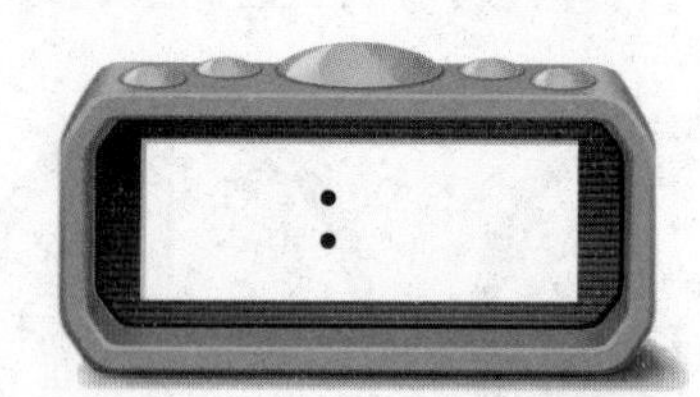

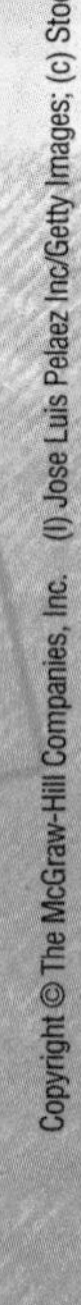

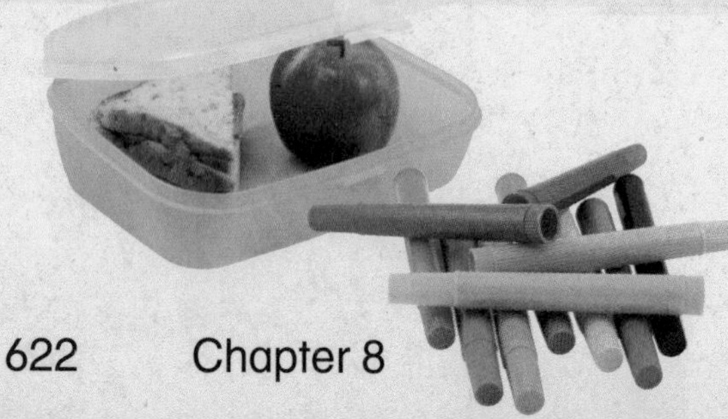

Now I Know!!!

Chapter 9

Two-Dimensional Shapes and Equal Shares

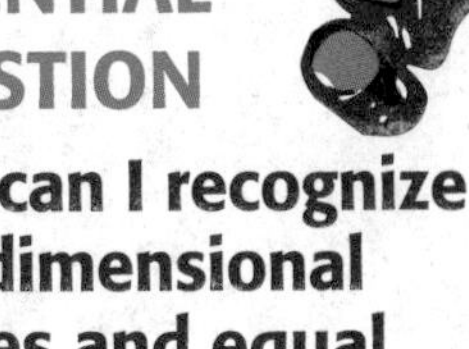

ESSENTIAL QUESTION

How can I recognize two-dimensional shapes and equal shares?

We're on the Farm!

Watch a video!

Watch

My Common Core State Standards

Geometry

1.G.1 Distinguish between defining attributes (e.g., triangles are closed and three-sided) versus non-defining attributes (e.g., color, orientation, overall size); build and draw shapes to possess defining attributes.

1.G.2 Compose two-dimensional shapes (rectangles, squares, trapezoids, triangles, half-circles, and quarter-circles) or three-dimensional shapes (cubes, right rectangular prisms, right circular cones, and right circular cylinders) to create a composite shape, and compose new shapes from the composite shape.

1.G.3 Partition circles and rectangles into two and four equal shares, describe the shares using the words *halves, fourths,* and *quarters,* and use the phrases *half of, fourth of,* and *quarter of.* Describe the whole as two of, or four of the shares. Understand for these examples that decomposing into more equal shares creates smaller shares.

1. Make sense of problems and persevere in solving them.
2. Reason abstractly and quantitatively.
3. Construct viable arguments and critique the reasoning of others.
4. Model with mathematics.
5. Use appropriate tools strategically.
6. Attend to precision.
7. Look for and make use of structure.
8. Look for and express regularity in repeated reasoning.

= focused on in this chapter

Name ....

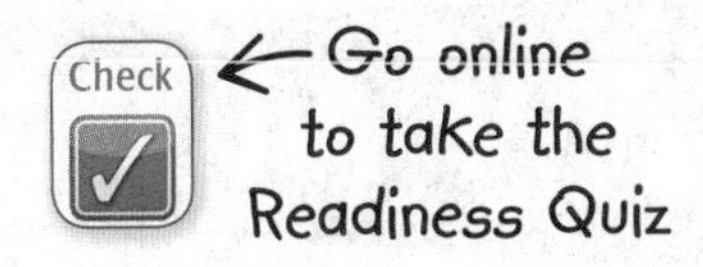

Draw lines to match the objects with the same shape.

1.  

2. 

3.  

4. JD made this picture frame in art class. What shape is it? Circle the name.

triangle   square   rectangle

5. Circle the shape that is different.

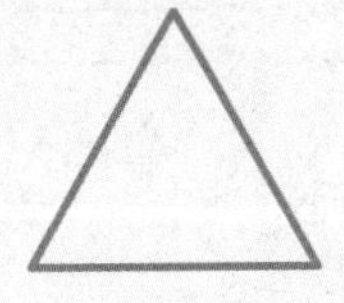  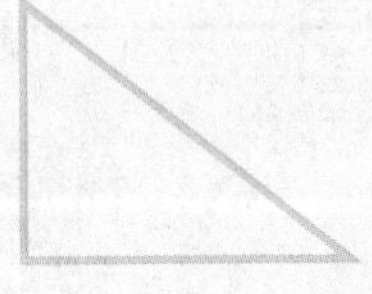

6. Circle the shapes that are the same.

 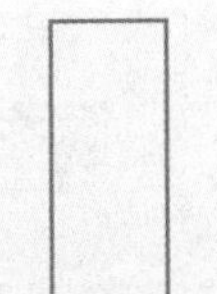 

How Did I Do? **Shade the boxes to show the problems you answered correctly.**

1	2	3	4	5	6

Name ....

My Math Words

Review Vocabulary

circle    square    triangle

Making Connections

Write the word that names each shape. Then draw an object in your classroom that is the same shape.

My Example

My Example

My Example

My Vocabulary Cards

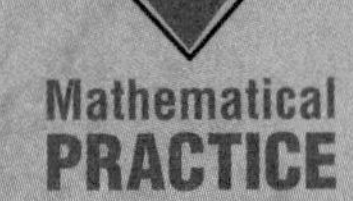

Mathematical
PRACTICE

Lesson 9-3

circle

Lesson 9-8

equal parts

2 equal parts

4 equal parts

Lesson 9-10

fourths

Lesson 9-9

halves

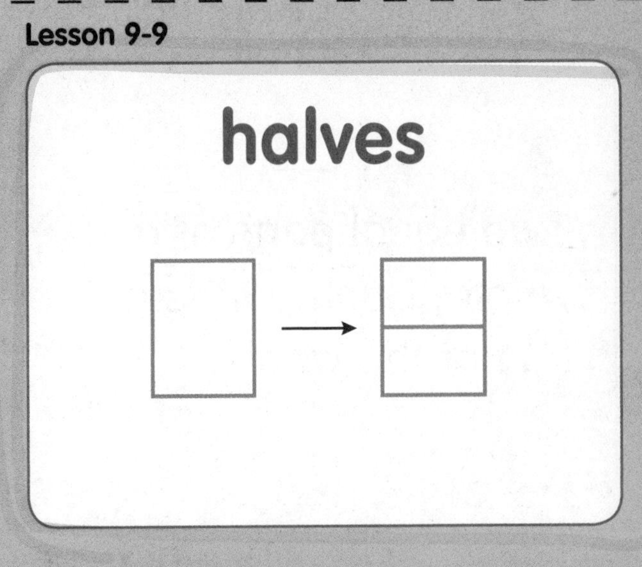

Lesson 9-1

rectangle

Lesson 9-1

side

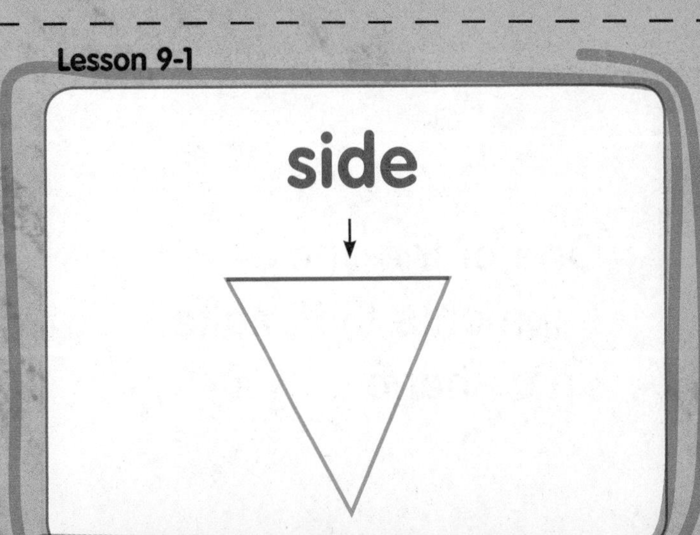

Teacher Directions:
Ideas for Use

- Tell students to create riddles for each word. Ask them to work with a friend to guess the word for each riddle.
- Have students sort the words by the number of letters in each word.

Parts of a whole that are the same size.

A closed round shape. Circles do not have sides and vertices.

Two equal parts of a whole. Each part is a half of the whole.

Four equal parts of a whole. Each part is a quarter or fourth of the whole.

One of the line segments that make up a shape.

A closed two-dimensional shape with four sides and four vertices. It has the same shape as a dollar bill.

My Vocabulary Cards

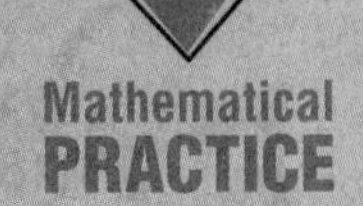

Mathematical PRACTICE

Lesson 9-1

square

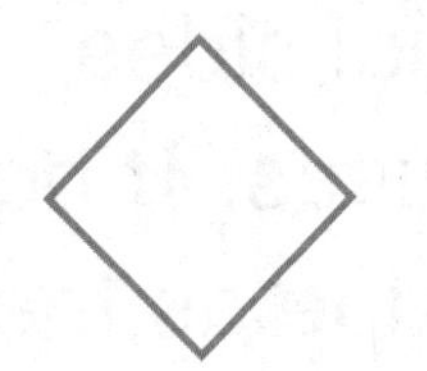

Lesson 9-2

trapezoid

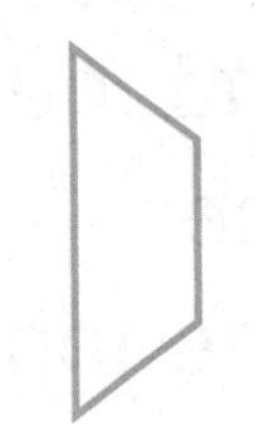
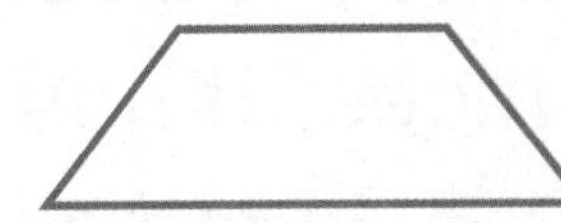

Lesson 9-2

triangle

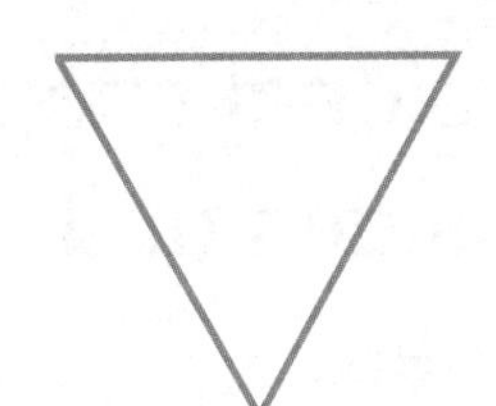
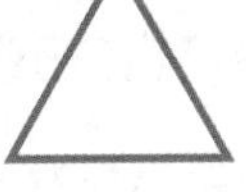

Lesson 9-1

two-dimensional shape

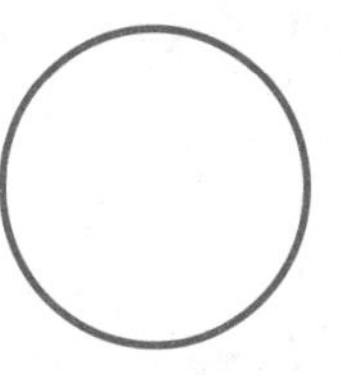
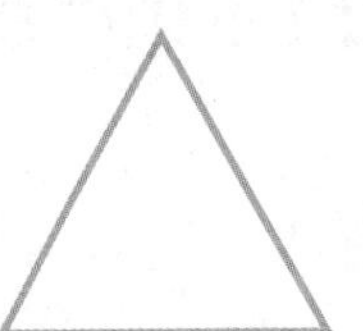

Lesson 9-1

vertex

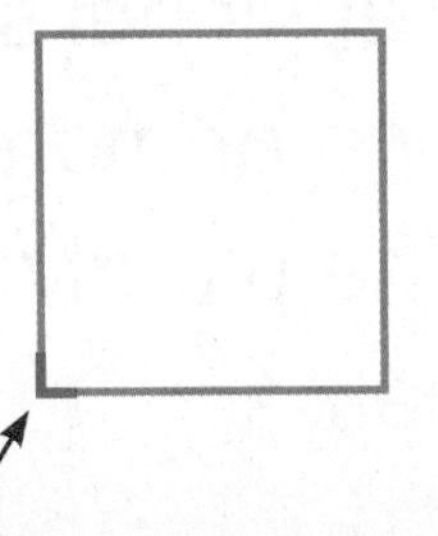

Lesson 9-8

whole

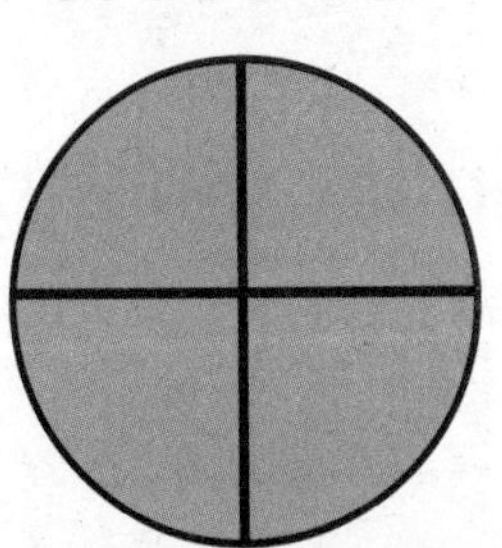

4 out of 4 parts shaded

Teacher Directions:
More Ideas for Use

- Have students write a tally mark on each card every time they read the word in this chapter or use it in their writing.
- Ask students to arrange cards to show an opposite pair. Have them explain the meaning of their pairing.

A closed two-dimensional shape with four sides and four vertices. It has the same shape as the top of a xylophone.

A closed two-dimensional shape with four equal sides and four vertices. It has the same shape as the side of a number cube.

A flat shape, such as a circle, a triangle, or a square.

A closed two-dimensional shape with three sides and three vertices.

The entire amount or all of the parts.

A point on a two-dimensional shape where two or more sides meet. The plural of vertex is vertices.

My Vocabulary Cards

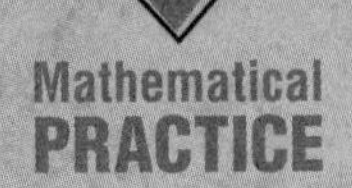

Lesson 9-5

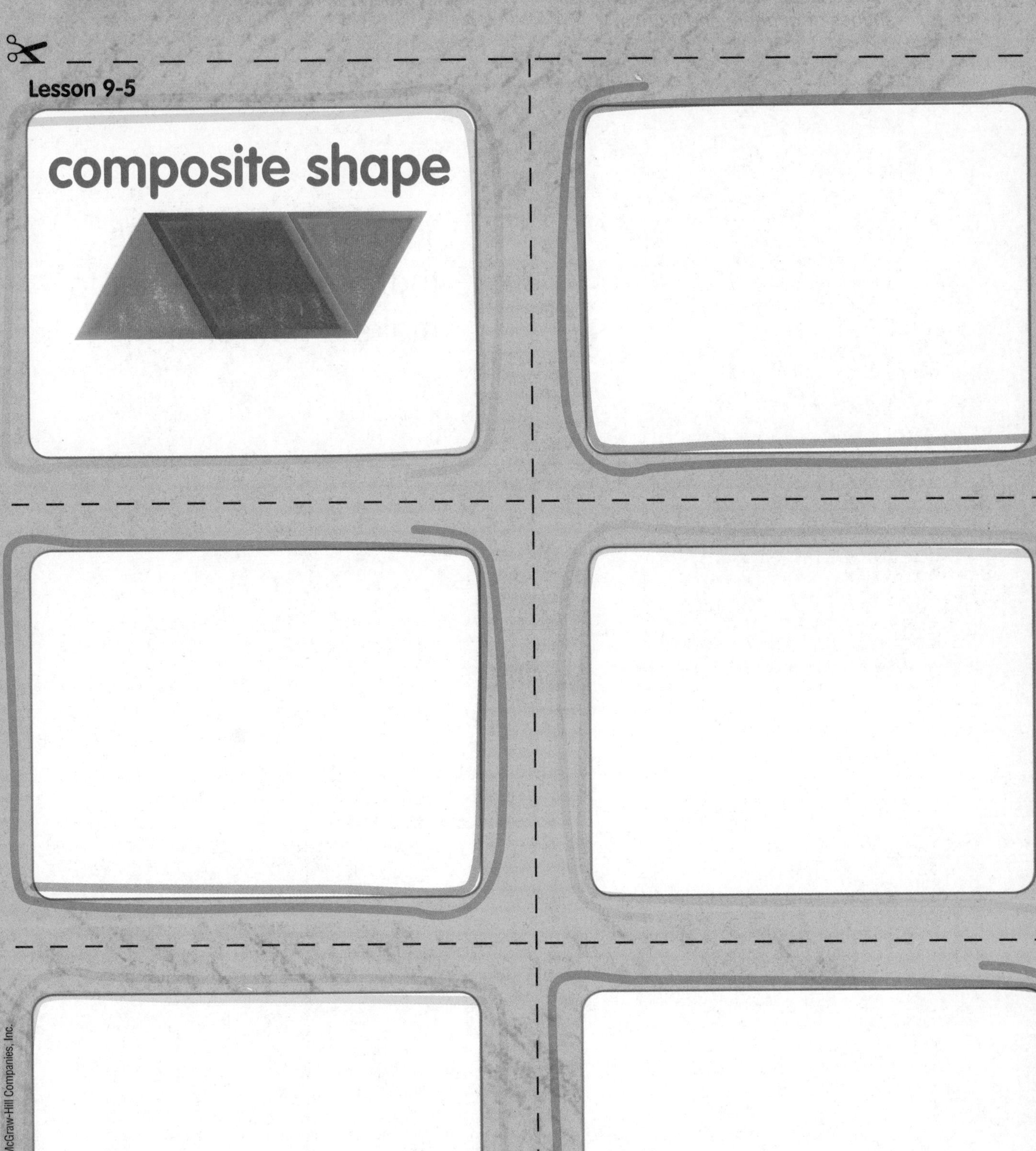

Teacher Directions:
More Ideas for Use

- Ask students to use the blank cards to draw or write words that will help them with concepts such as *make composite shapes* or *quarters and fourths.*
- Have students use the blank cards to write a word from a previous chapter that they would like to review.

Two or more shapes that are put together to make a new shape.

My Foldable

FOLDABLES® Follow the steps on the back to make your Foldable.

circle

_____ sides

_____ vertices

rectangle

_____ sides

_____ vertices

1
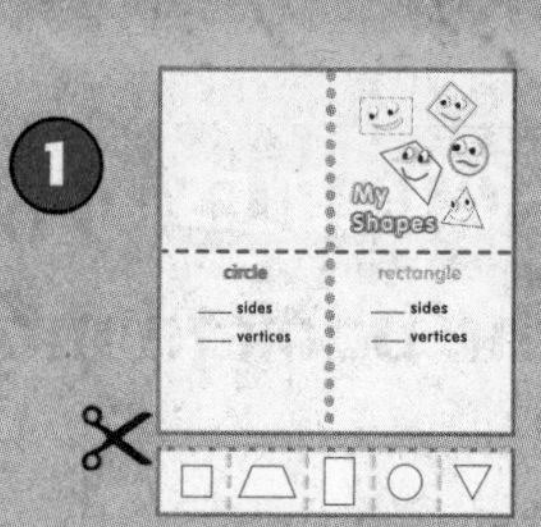

2

3

4
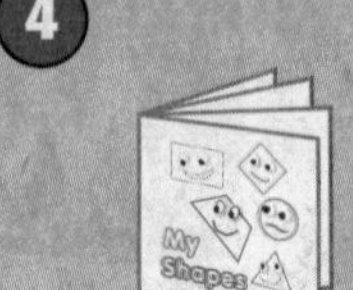

square

____ sides

____ vertices

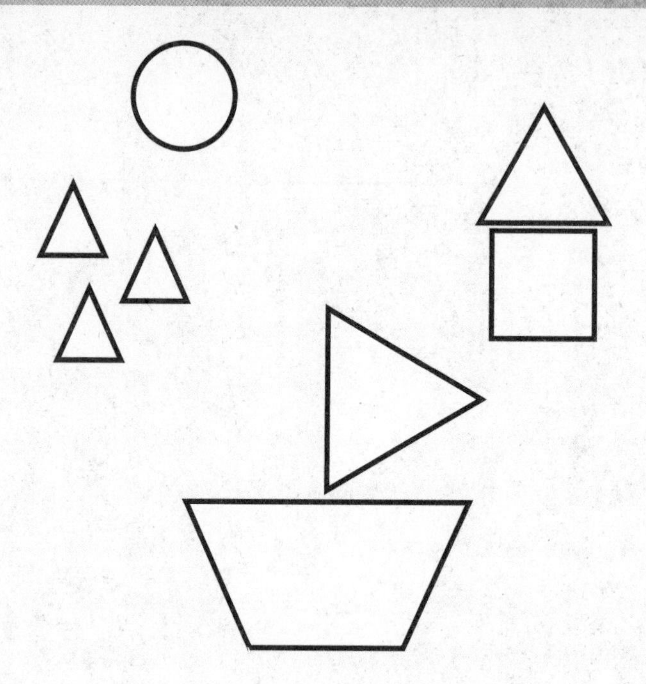

trapezoid

____ sides

____ vertices

triangle

____ sides

____ vertices

Name ....

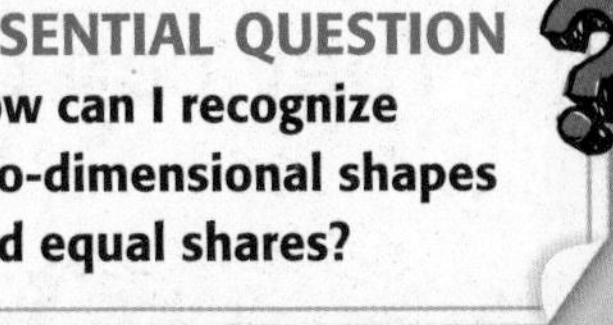

Geometry

1.G.1

CCSS

Squares and Rectangles

Lesson 1

ESSENTIAL QUESTION

How can I recognize two-dimensional shapes and equal shares?

Explore and Explain

_____ squares _____ rectangles

Teacher Directions: Use square and rectangle attribute blocks to make a farm picture. Trace the squares using a red crayon. Trace the rectangles using a blue crayon. Count the shapes. Write how many there are of each shape.

See and Show

Two-dimensional shapes are flat shapes. They can be open or closed.

Open

Closed

Squares and **rectangles** are two-dimensional shapes. They are closed. They have straight **sides** and vertices (**vertex**).

square

vertex

4 sides

4 vertices

rectangle

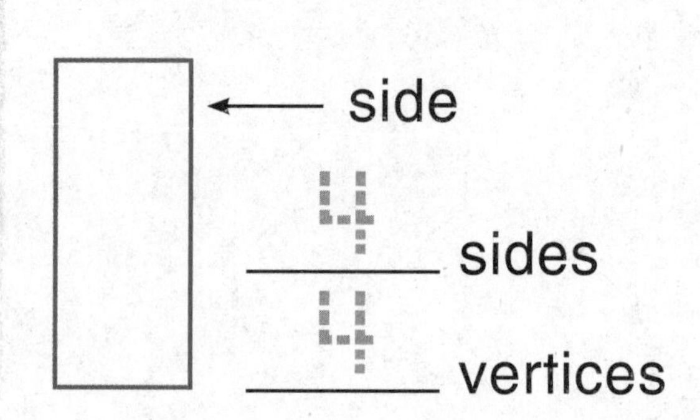

Write how many sides and vertices.

1.

_____ sides

_____ vertices

2.

_____ sides

_____ vertices

Talk Math How are a rectangle and a square alike?

Name ....

On My Own

Write how many sides and vertices.

3. 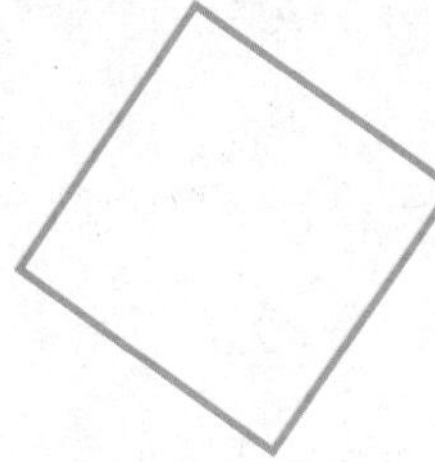

_____ sides

_____ vertices

4. 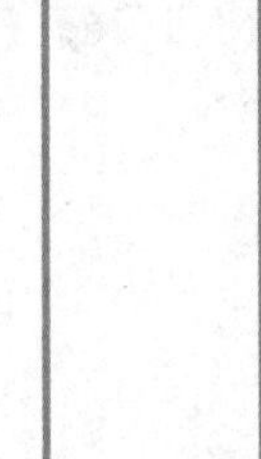

_____ sides

_____ vertices

5. 

_____ sides

_____ vertices

6. 

_____ sides

_____ vertices

Circle the objects that match the first shape.

7.   

8.    

Circle the closed shapes.

9.  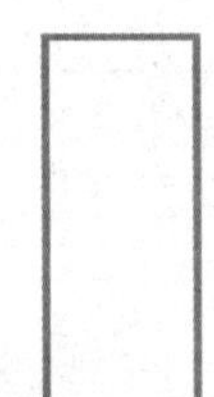 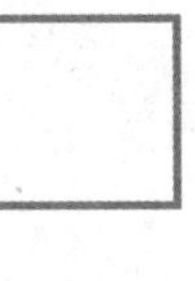   

Problem Solving

Name and draw each shape.

10. I am a two-dimensional shape that has 4 vertices. All of my sides are the same length. What shape am I?

____

11. I am a two-dimensional shape that has 4 sides. Two of my sides are long. Two of my sides are short. What shape am I?

____

Write Math Use the words *closed*, *sides*, and *vertices* to describe the square.

____

____

____

Name ....

My Homework

Lesson 1
Squares and Rectangles

Homework Helper

Need help? connectED.mcgraw-hill.com

Squares and rectangles are closed two-dimensional shapes. They have straight sides and vertices.

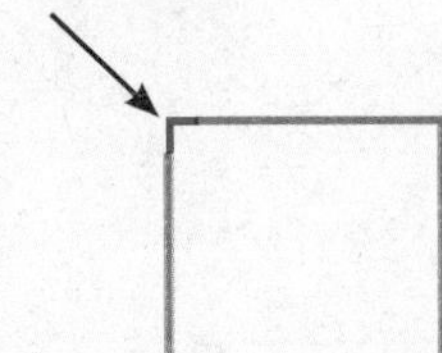

4 sides
4 vertices

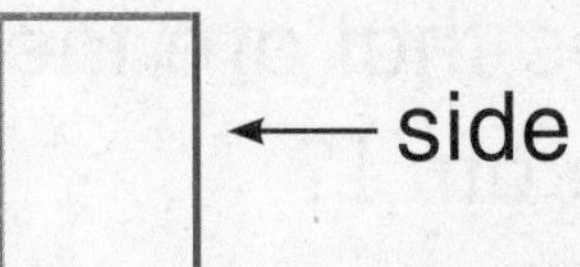

4 sides
4 vertices

Practice

Write how many sides and vertices.

1. 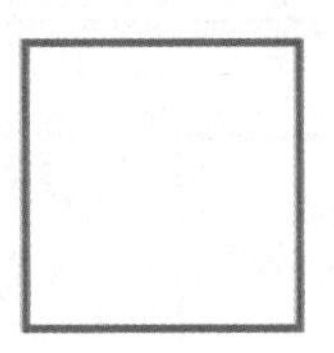

_____ sides

_____ vertices

2. 

_____ sides

_____ vertices

3. 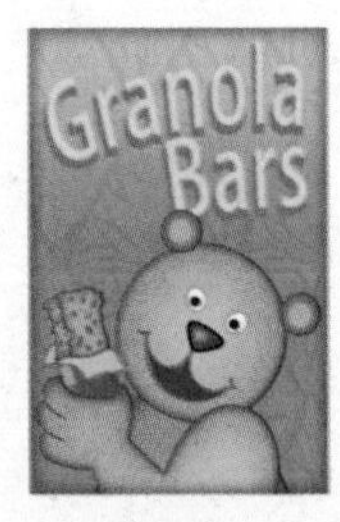

_____ sides

_____ vertices

4. 

_____ sides

_____ vertices

Count and write how many squares and rectangles you see in the robot.

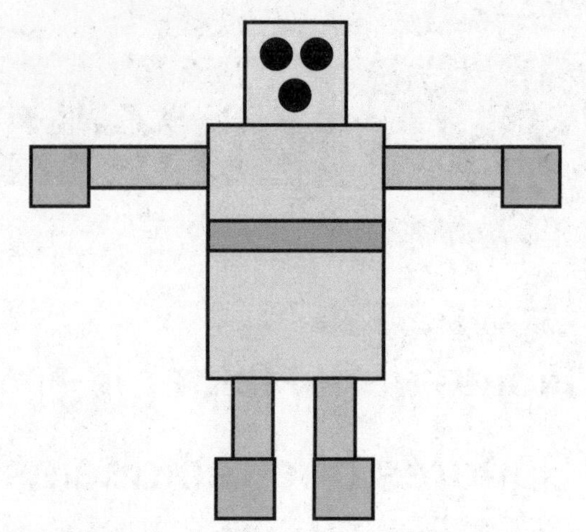

5. _______ squares

6. _______ rectangles

Draw and write the name of the shape.

7. I am a two-dimensional shape that has 4 sides that are the same length. What shape am I?

____

Vocabulary Check 

Draw lines to match.

8. **rectangle**

9. **square**

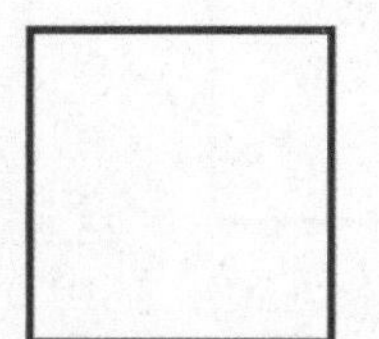

Math at Home Ask your child to draw a picture using only squares and rectangles.

Name ______________________

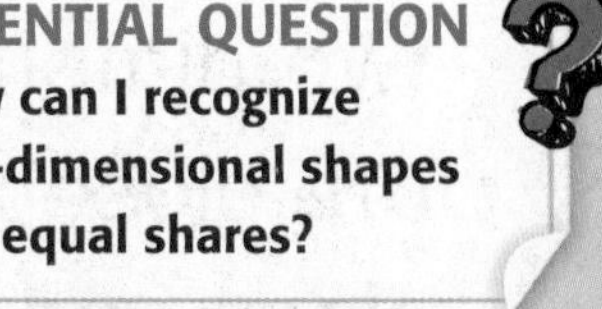

Triangles and Trapezoids

Lesson 2

ESSENTIAL QUESTION
How can I recognize two-dimensional shapes and equal shares?

Explore and Explain

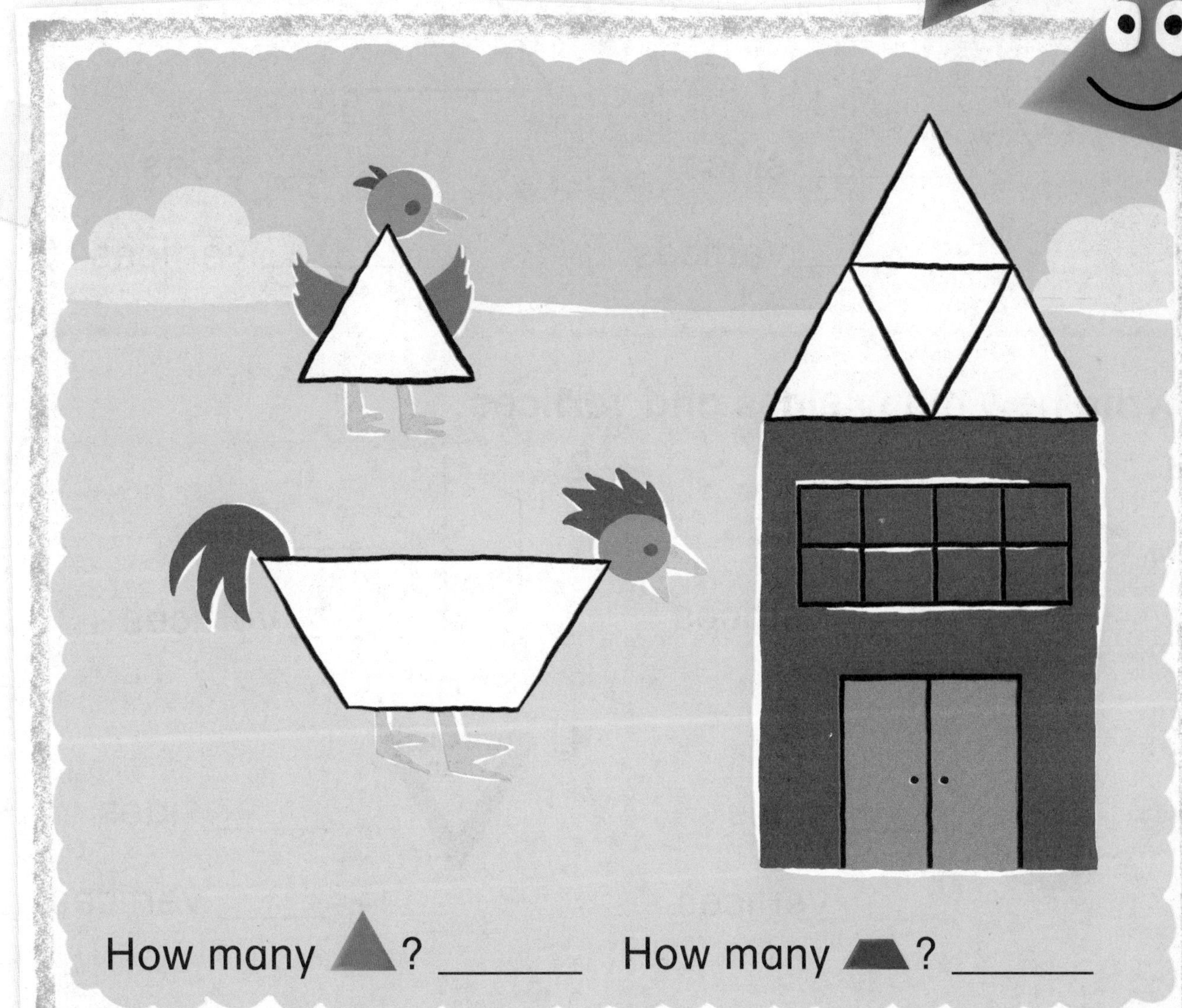

How many ▲? ______ How many ⏢? ______

Teacher Directions: Use trapezoid and triangle pattern blocks. Find the white shapes in the picture. Color the trapezoids red. Color the triangles green. Write how many of each shape you colored. Describe the shapes.

See and Show

Mathematical **PRACTICE**

Triangles and **trapezoids** are two-dimensional shapes. They are closed. They have straight sides and vertices.

triangle

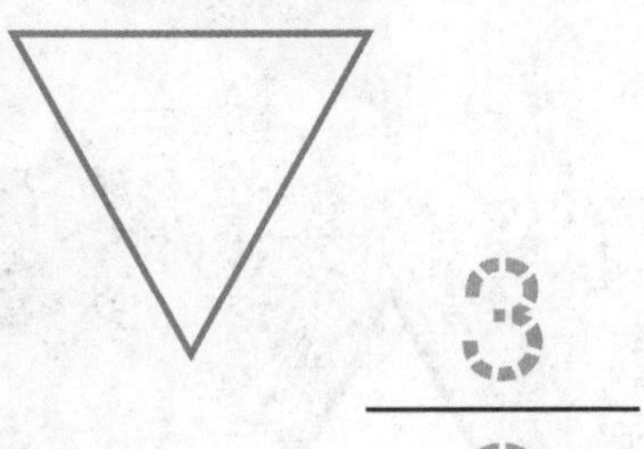

3 sides

3 vertices

trapezoid

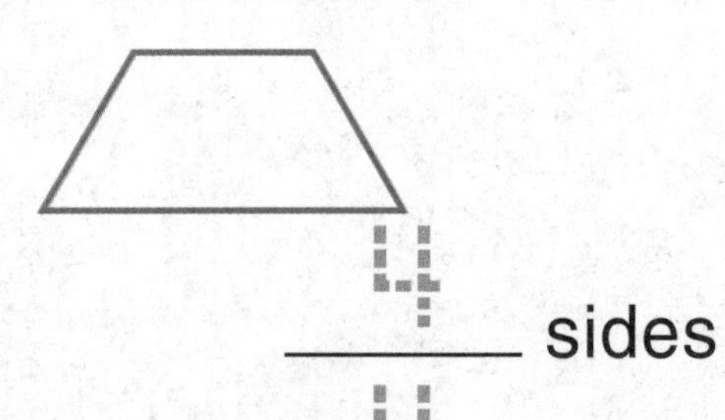

4 sides

4 vertices

Write how many sides and vertices.

1. 

_____ sides

_____ vertices

2. 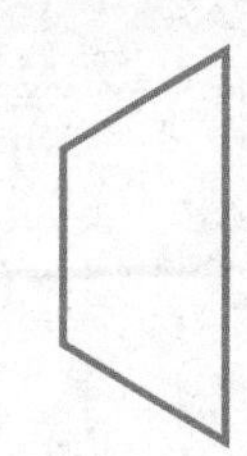

_____ sides

_____ vertices

3. 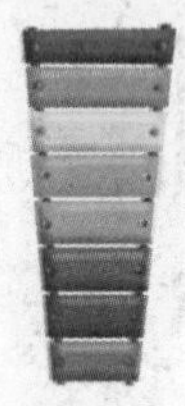

_____ sides

_____ vertices

4. 

_____ sides

_____ vertices

Talk Math How are a triangle and a trapezoid different?

Name ....

On My Own

Write how many sides and vertices.

5. _____ sides

_____ vertices

6. 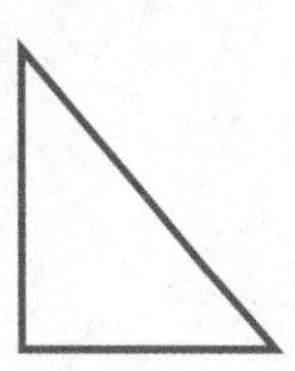

_____ sides

_____ vertices

Circle the objects that match the description.

7. 4 sides
4 vertices

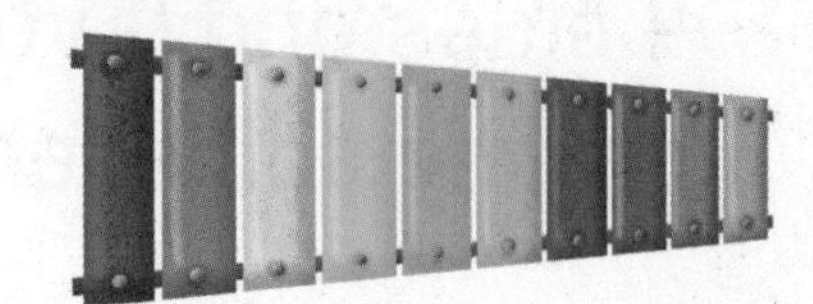

8. 3 sides
3 vertices

Circle the closed shapes.

9.

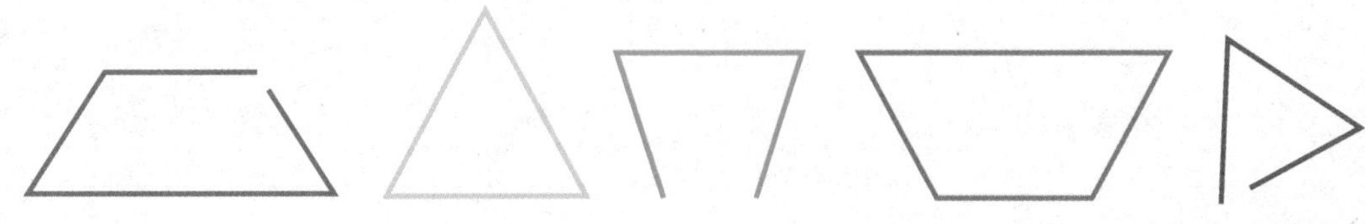

Problem Solving

Mathematical PRACTICE

Draw and write the name of each shape.

10. I am a two-dimensional shape that has 3 sides and 3 vertices. What shape am I?

____

11. I am a two-dimensional shape that has 4 sides and 4 vertices. Only 2 of my sides are the same length. What shape am I?

____

HOT Problem Ty has 3 different shapes. The shapes have 11 sides and 11 vertices in all. What shapes can he have? Explain.

Name ....

My Homework

Lesson 2
Triangles and Trapezoids

Homework Helper

eHelp Need help? connectED.mcgraw-hill.com

Triangles and trapezoids are closed two-dimensional shapes. They have vertices and straight sides.

triangle

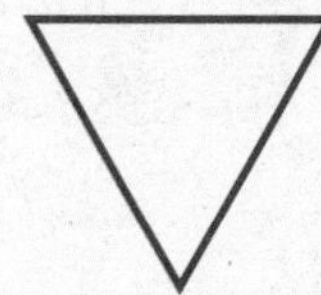

3 sides
3 vertices

trapezoid

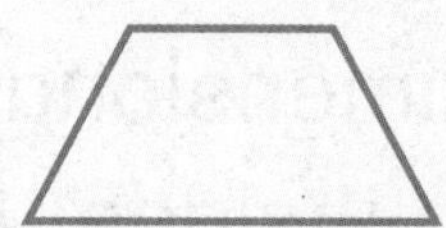

4 sides
4 vertices

Practice

Write how many sides and vertices.

1.

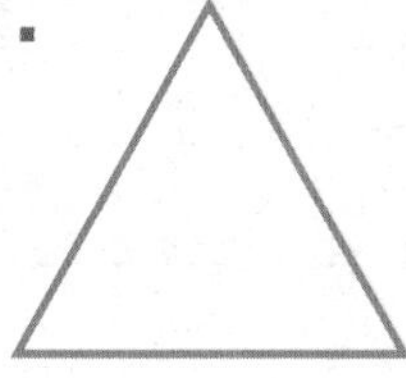

_____ sides

_____ vertices

2.

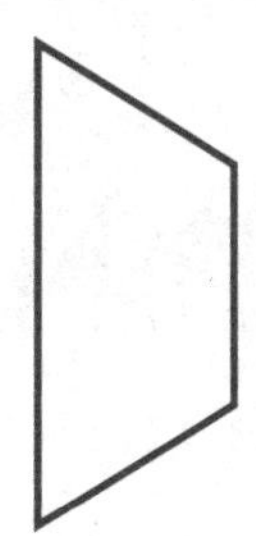

_____ sides

_____ vertices

3.

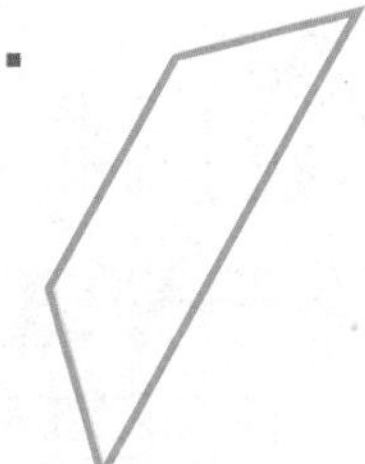

_____ sides

_____ vertices

4.

_____ sides

_____ vertices

5. Color all of the triangles red. Color all of the trapezoids purple. Then write how many.

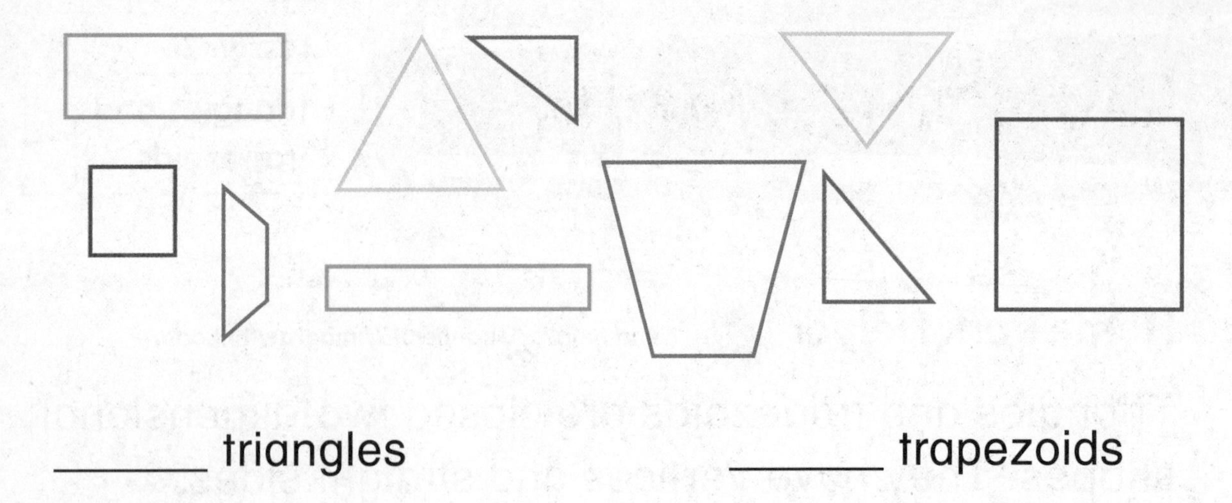

______ triangles ______ trapezoids

Draw and write the name of the shape.

6. I am a two-dimensional shape that has less than 4 sides. All of my sides are straight. What shape am I?

____

Vocabulary Check

Draw lines to match.

7. **triangle**

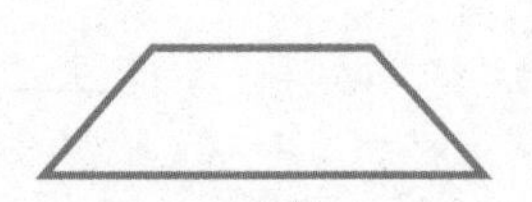

8. **trapezoid**

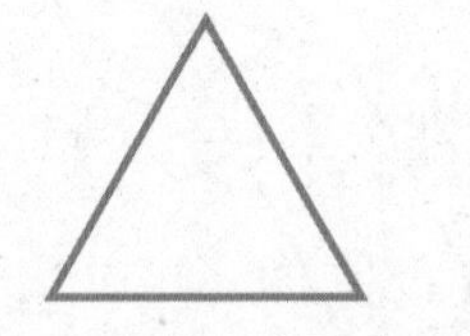

Math at Home Have your child compare a triangle and a trapezoid using words such as sides and vertices.

Name

Circles

Lesson 3

ESSENTIAL QUESTION
How can I recognize two-dimensional shapes and equal shares?

Explore and Explain

Around we go!

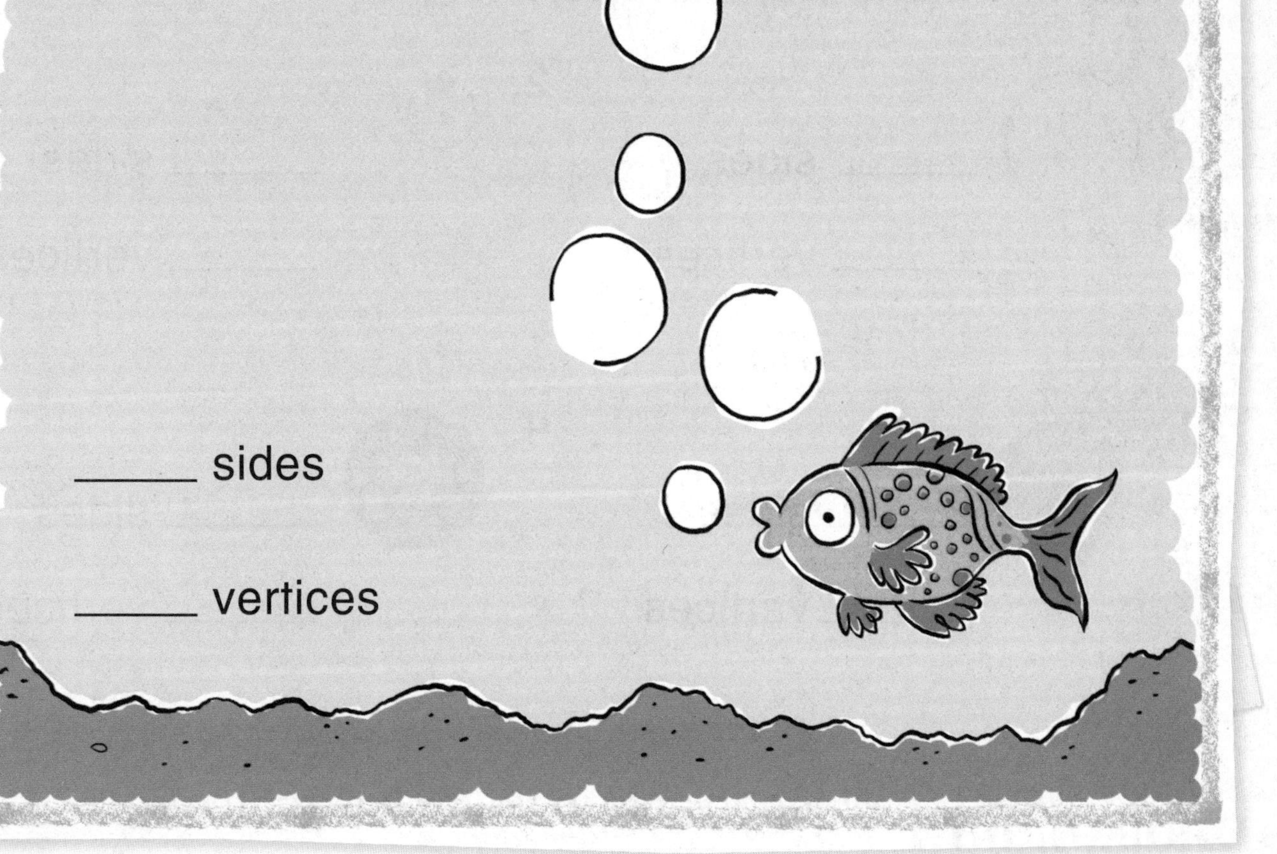

Teacher Directions: Fish blow bubbles in the shape of circles. Finish drawing the bubbles the fish blew. Then use circle attribute blocks to trace and draw 4 more bubbles. Describe the shapes. Write how many sides and vertices.

See and Show

Mathematical PRACTICE

Circles are two-dimensional shapes. They are closed and round. They do not have sides and vertices.

circle

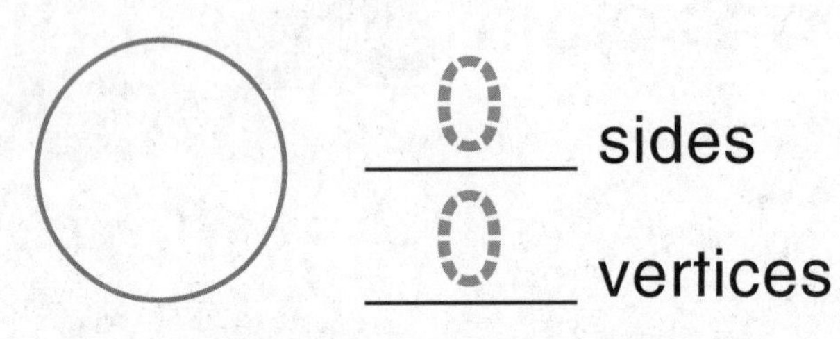

0 sides

0 vertices

Write how many sides and vertices.

1. ______ sides

   ______ vertices

2. 

   ______ sides

   ______ vertices

3. 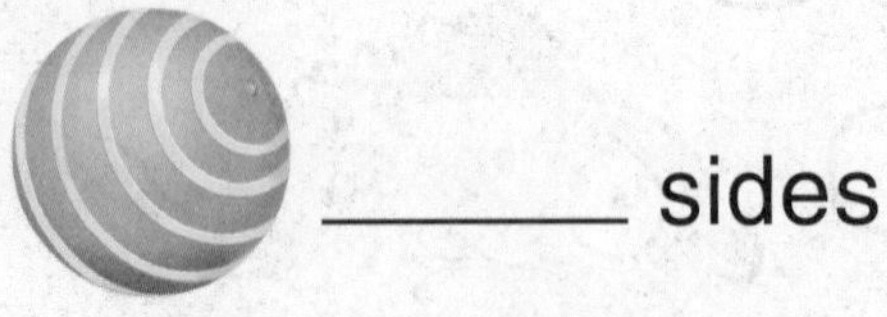

   ______ sides

   ______ vertices

4. 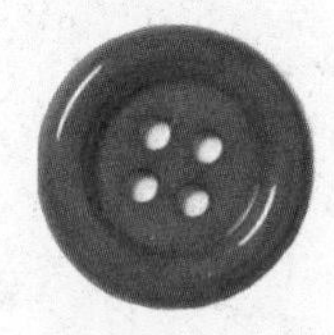

   ______ sides

   ______ vertices

Talk Math What objects in your classroom are the shape of a circle?

Name

On My Own

Write how many sides and vertices.

5. 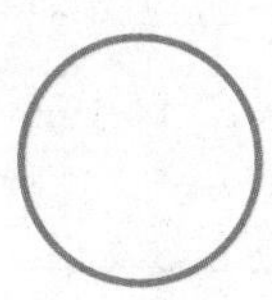 ______ sides ______ vertices

6. 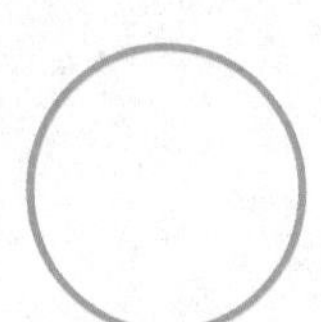 ______ sides ______ vertices

7. 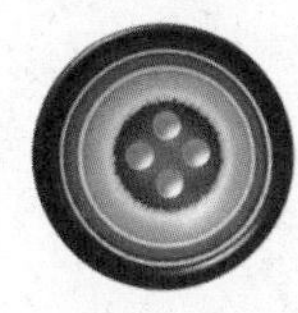 ______ sides ______ vertices

8.  ______ sides ______ vertices

Circle the objects that match the first shape.

9. 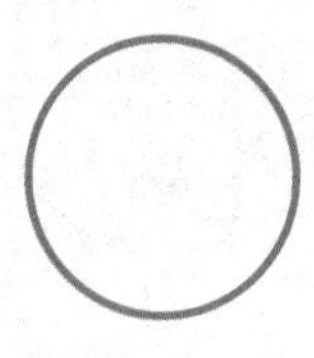 

Circle the closed shapes.

10. 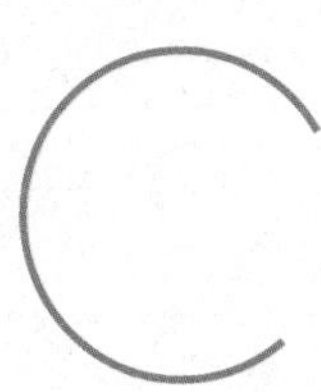  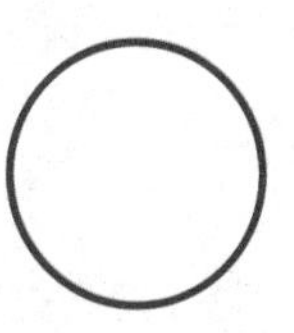  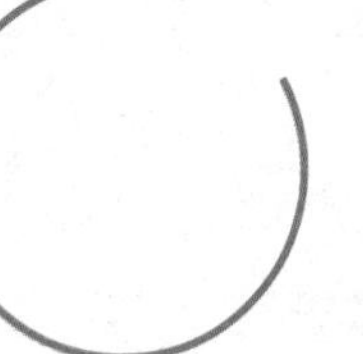   

I swim in circles!

Problem Solving

11. I am a two-dimensional shape that has no sides or vertices. Write the name of the shape. Draw a picture of the shape.

____

12. Draw a picture of a two-dimensional shape. Write how many sides. Write how many vertices.

_____ sides

_____ vertices

HOT Problem Majid described this shape as a two-dimensional shape with 4 sides. Majid is wrong. Make it right.

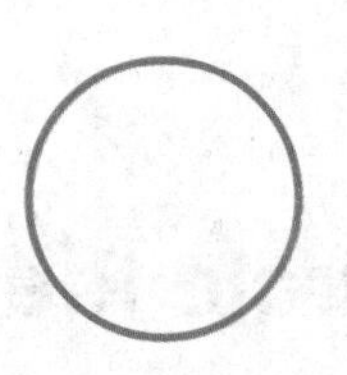

Name ....

My Homework

Lesson 3
Circles

Homework Helper

Need help? connectED.mcgraw-hill.com

Circles are closed round shapes. They do not have sides and vertices.

circle

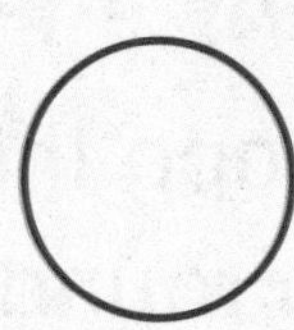

0 sides
0 vertices

Practice

Write how many sides and vertices.

1. 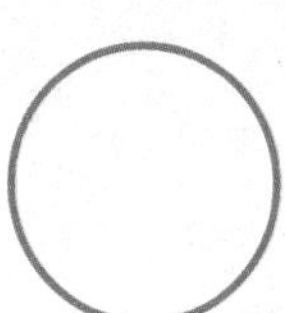
_____ sides
_____ vertices

2. 
_____ sides
_____ vertices

3. 
_____ sides
_____ vertices

4. 
_____ sides
_____ vertices

5. 
_____ sides
_____ vertices

6. 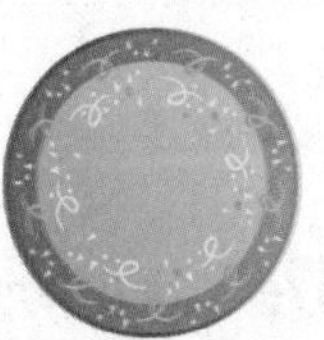
_____ sides
_____ vertices

7. Amad drew the shapes below.

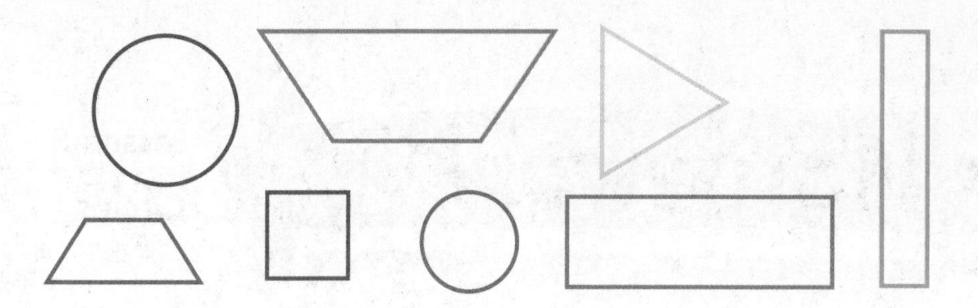

How many shapes have 4 sides? ______ shapes

How many shapes have 0 vertices? ______ shapes

Draw and write the name of the shape.

8. I am a two-dimensional shape that has no straight sides. I am round. What shape am I?

____

Vocabulary Check

Circle the shape that shows the vocabulary word.

9. **circle**

Math at Home Cut several different sized circles out of paper. Have your child create a picture by gluing the circles onto another sheet of paper.

Name

Compare Shapes

Lesson 4

ESSENTIAL QUESTION
How can I recognize two-dimensional shapes and equal shares?

Explore and Explain

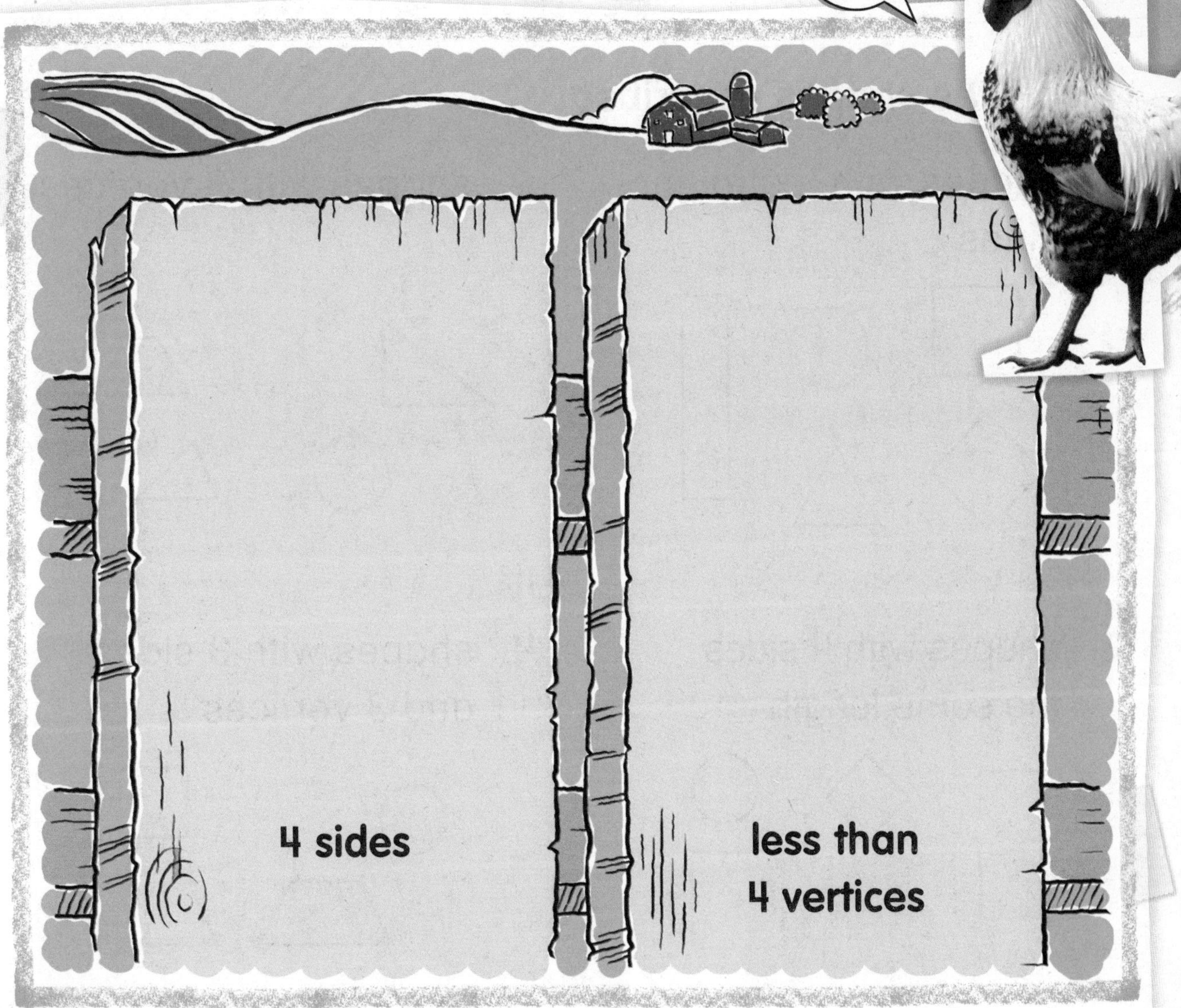

Teacher Directions: Use attribute block circles, squares, rectangles, and triangles. Place the shapes that have 4 sides on the left fence post. Trace the shapes. Place the shapes that have less than 4 vertices on the right fence post. Trace the shapes.

See and Show

Mathematical PRACTICE

You can compare and sort two-dimensional shapes.

Circle the shapes that have straight sides.

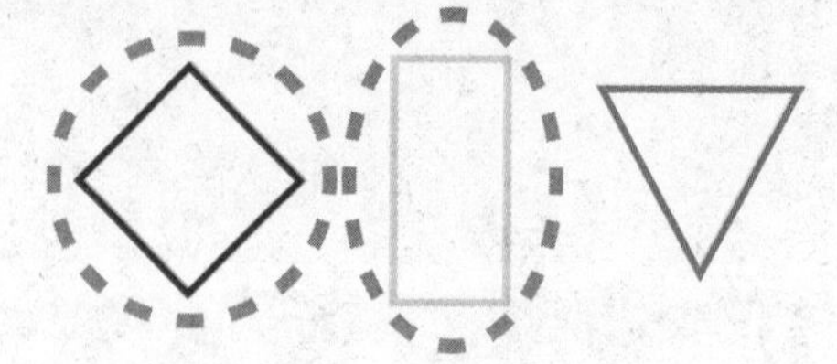

Circle the shapes with more than 3 vertices.

Circle the shapes described.

1. shapes with 4 straight sides

2. shapes with 3 vertices

3. shapes with 4 sides the same length

4. shapes with 3 sides and 3 vertices

Talk Math How do you compare two-dimensional shapes?

Name ....

What shape am I?

On My Own

Circle the shapes described.

5. shapes with 0 vertices

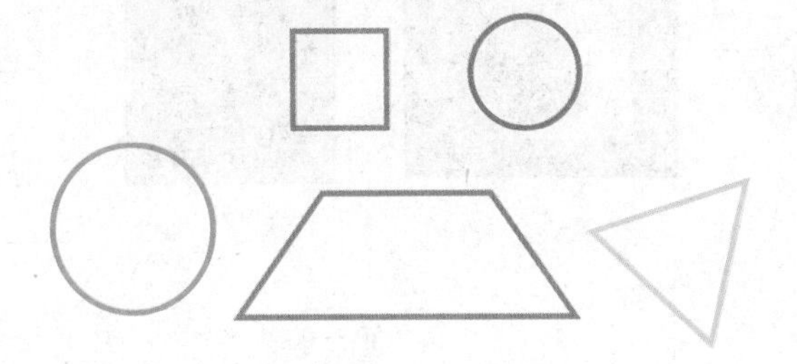

6. shapes with 4 sides

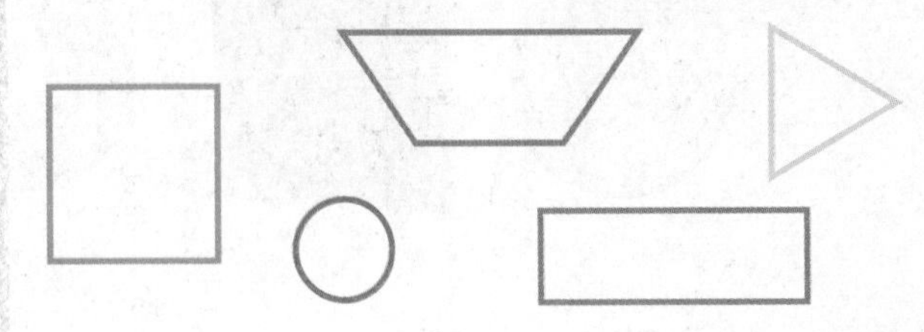

7. shapes with straight sides

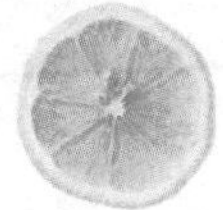

8. shapes with 0 vertices

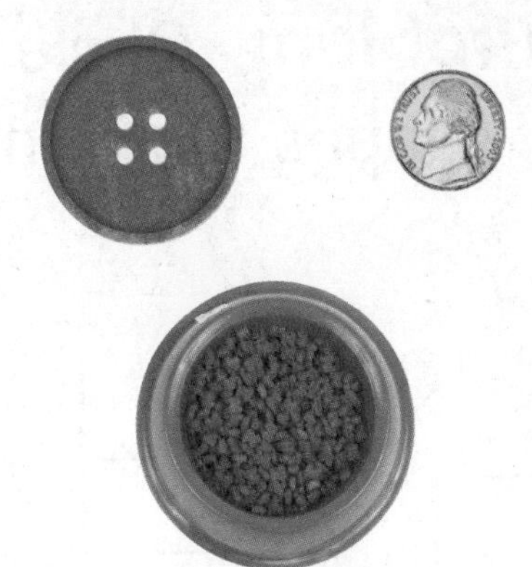

9. shapes with 0 straight sides

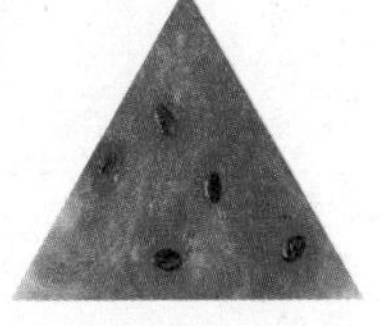

10. shapes that are not curved

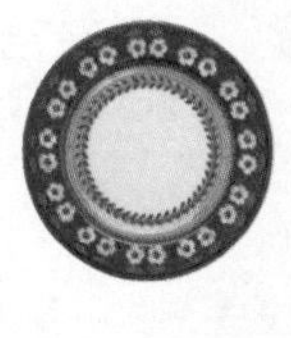

Problem Solving

11. Madeline sees these objects in her school. How many of the objects have more than three sides?

_____ objects

HOT Problem Circle all of the same type of shapes. Explain.

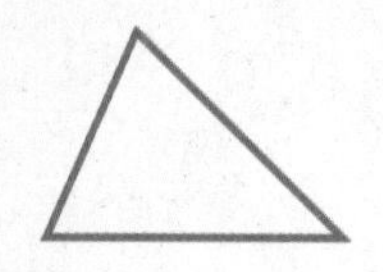

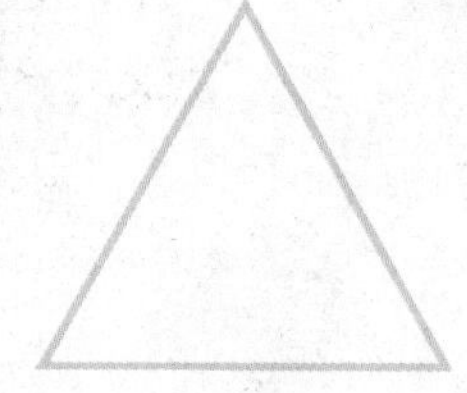
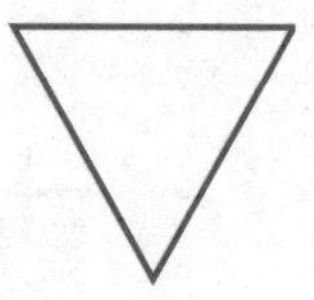

Name ______________________

My Homework

Lesson 4
Compare Shapes

Homework Helper

Need help? connectED.mcgraw-hill.com

You can compare and sort two-dimensional shapes.

shapes that have 4 straight sides

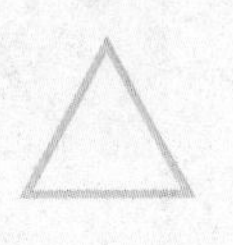

shapes that are closed and have 3 vertices

Practice

Circle the shapes described.

1. shapes with 0 vertices

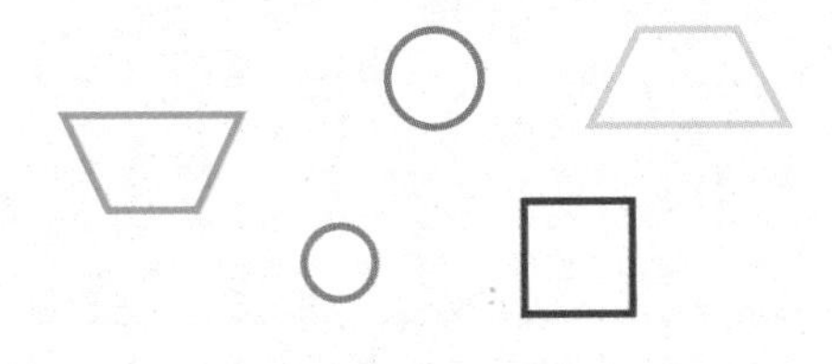

2. shapes with 3 sides

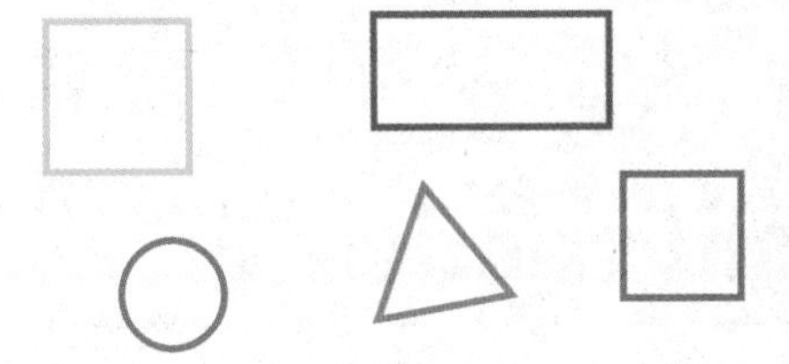

3. shapes with more than 2 sides

4. shapes that are closed

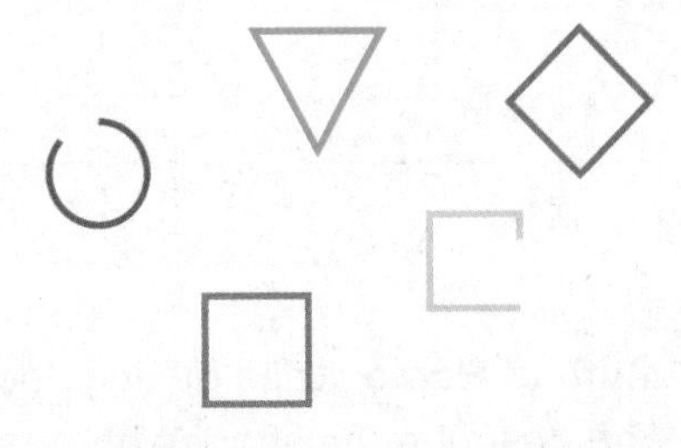

Circle the shapes described.

5. shapes with more than 2 straight sides

6. shapes with less than 4 vertices

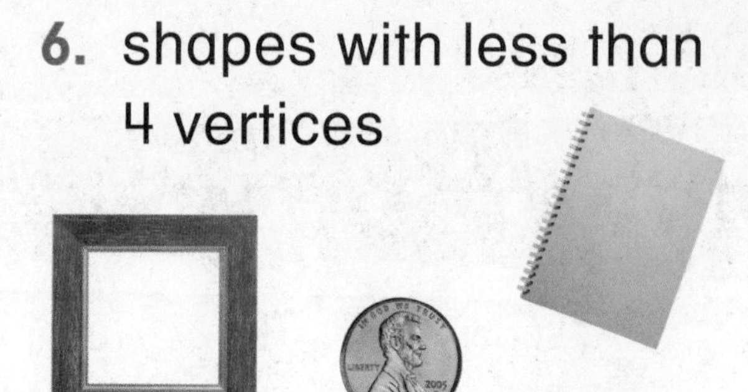

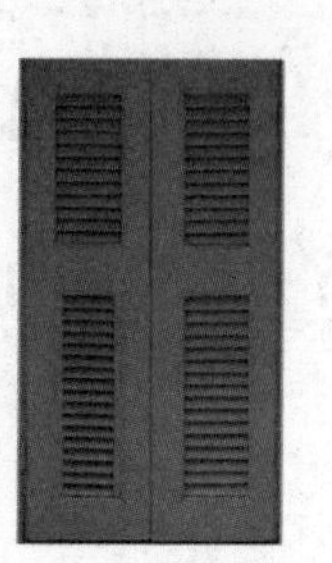

7. Draw a shape that has 4 vertices and 2 pair of sides that are different lengths.

Test Practice

8. Which shape has 3 sides and 3 vertices?

circle ◯ square ◯ triangle ◯ rectangle ◯

Math at Home While driving, look at road signs together. Ask your child to name and describe the shapes he or she sees.

Name ....

Check My Progress

Vocabulary Check 

Draw lines to match.

1. **rectangle** 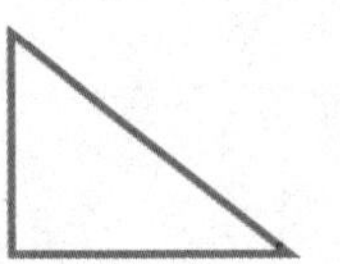

2. **square**  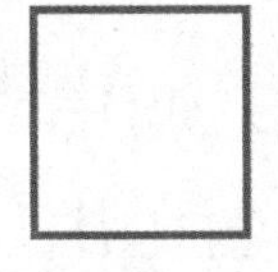

3. **trapezoid** 

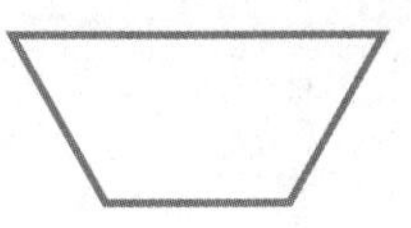

4. **triangle** 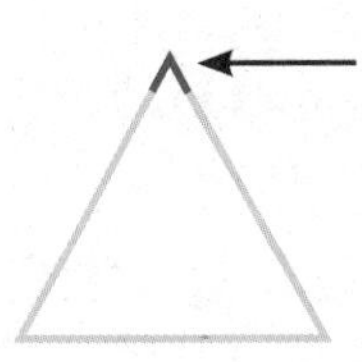

5. **vertex**

Concept Check

Write how many sides and vertices.

6. 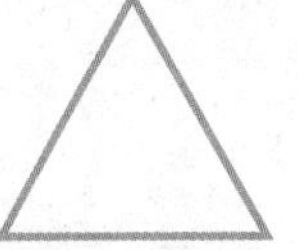

_____ sides

_____ vertices

7. 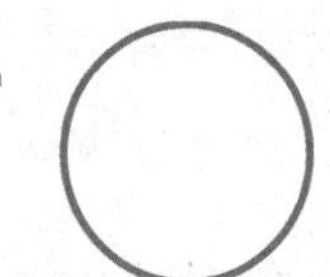

_____ sides

_____ vertices

Write how many sides and vertices.

8. _____ sides

_____ vertices

9. _____ sides

_____ vertices

Circle the shapes described.

10. shapes with 3 sides

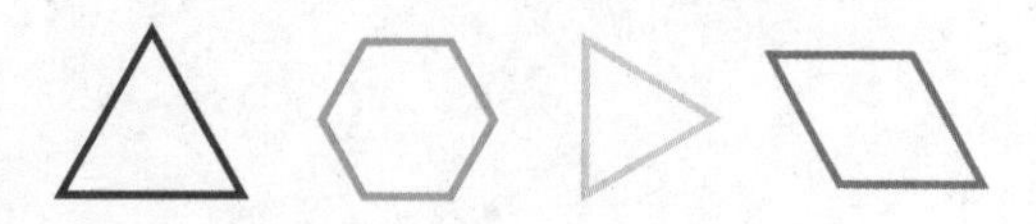

11. shapes with 0 sides

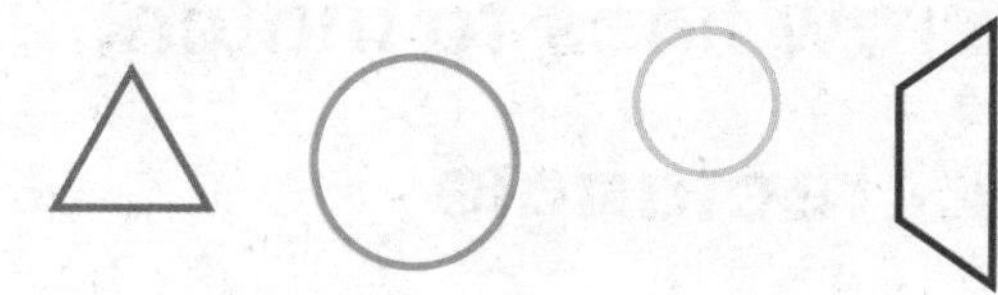

12. shapes with 4 sides the same length

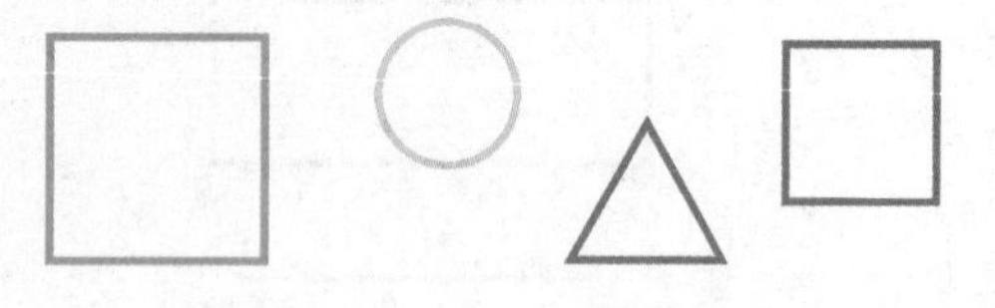

13. shapes with 4 sides and 4 vertices

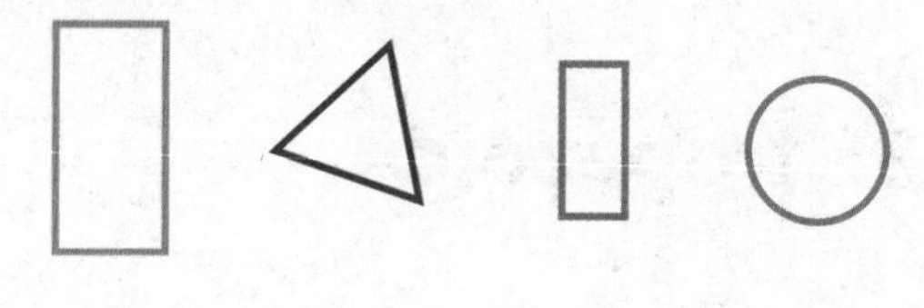

Test Practice

14. Jill has a shape with 4 sides of equal length and 4 vertices. Carlos has a shape with no sides or vertices. What are the two shapes?

square, circle ○	square, triangle ○
rectangle, square ○	triangle, circle ○

Name

Composite Shapes

Lesson 5

ESSENTIAL QUESTION
How can I recognize two-dimensional shapes and equal shares?

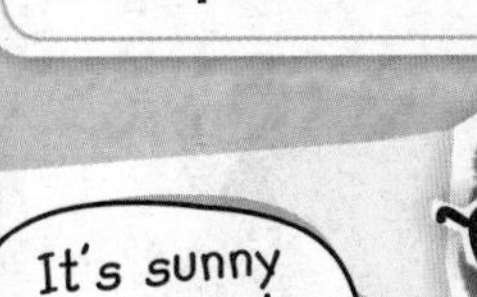

Explore and Explain

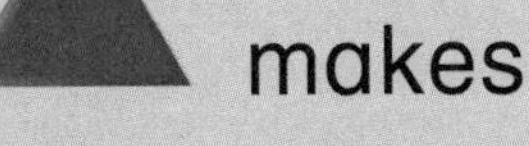
______ makes

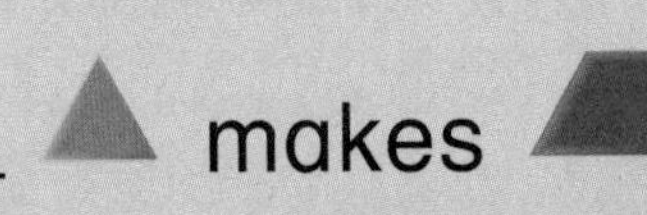
______ makes

Teacher Directions: Use pattern blocks to make each new shape. Trace the pattern blocks to show your work. Write how many pattern blocks you used.

See and Show

You can put together shapes to make a new shape.

This new shape is called a **composite shape**.

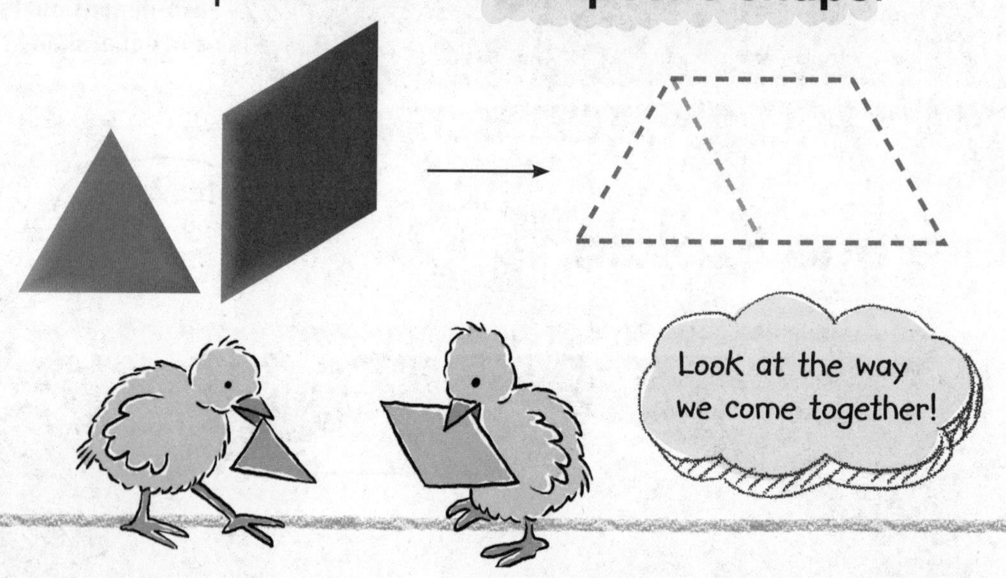

Use two pattern blocks to make each shape. Draw a line to show your model. Circle the blocks you use.

1.

2.

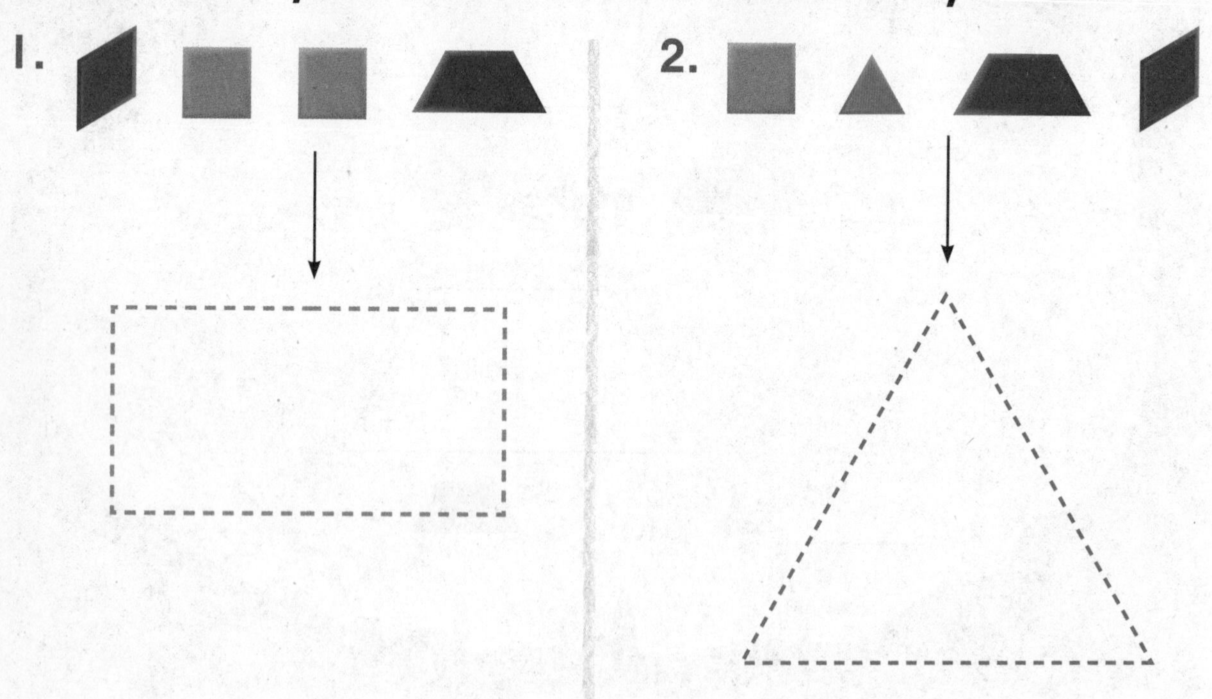

Talk Math How can you find which shapes are needed to make composite shapes?

Name

On My Own

Use two pattern blocks to make each shape. Draw a line to show your model. Circle the blocks you use.

3. 4.

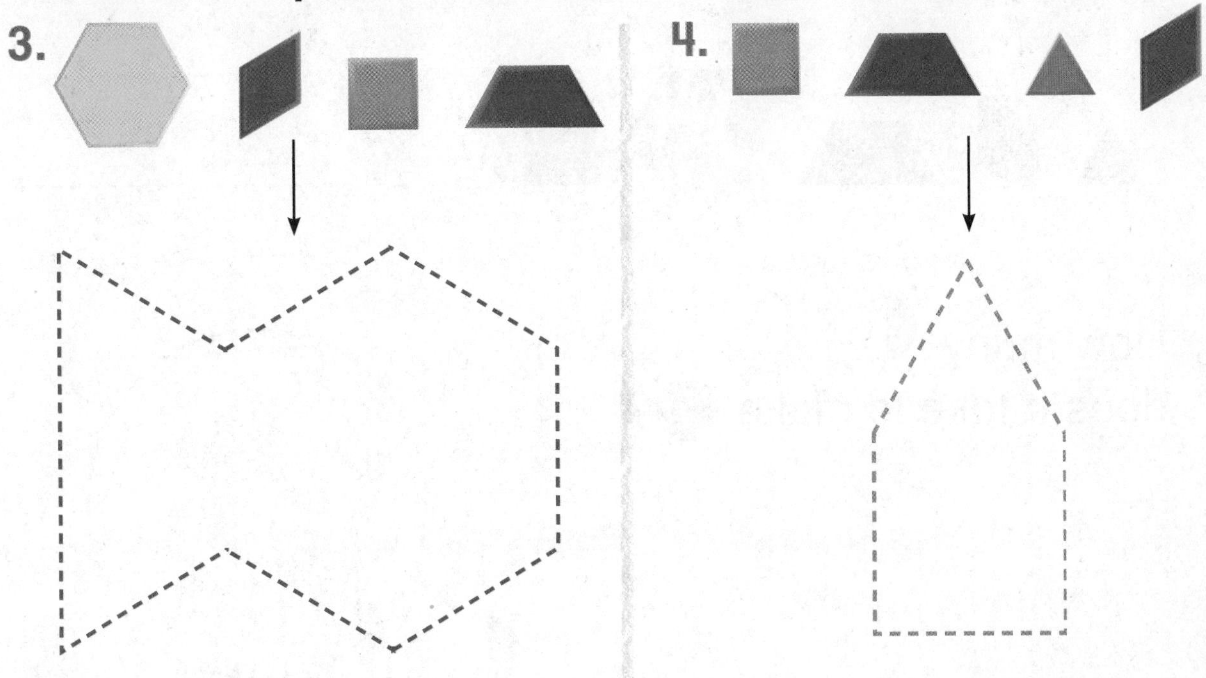

Pick two of the pattern blocks shown to make a composite shape. Draw the shape. Circle the blocks you use.

5.

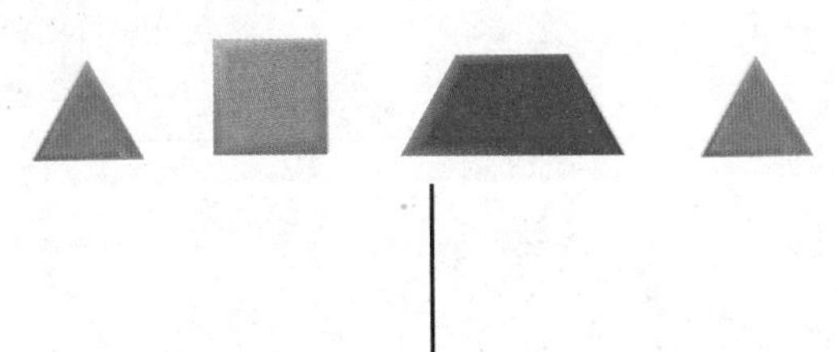

6.

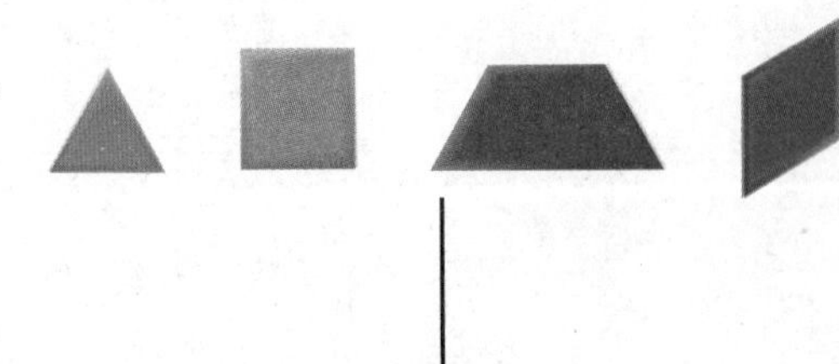

Problem Solving

Mathematical PRACTICE

Answer the questions. Draw lines to show your work.

7. Circle the pattern block that can be used 2 times to make this shape.

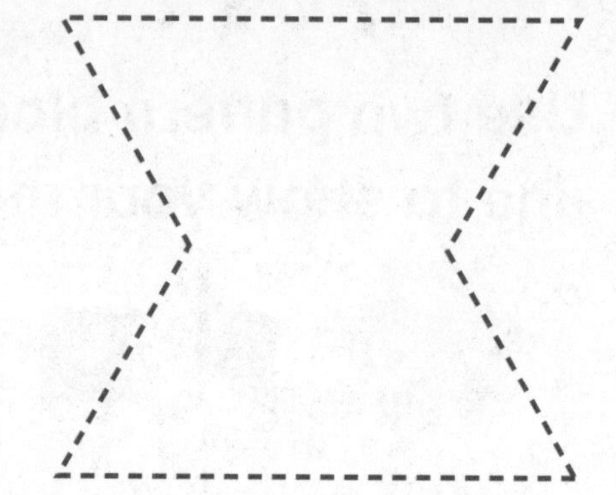

8. How many ◗ does it take to make ●?

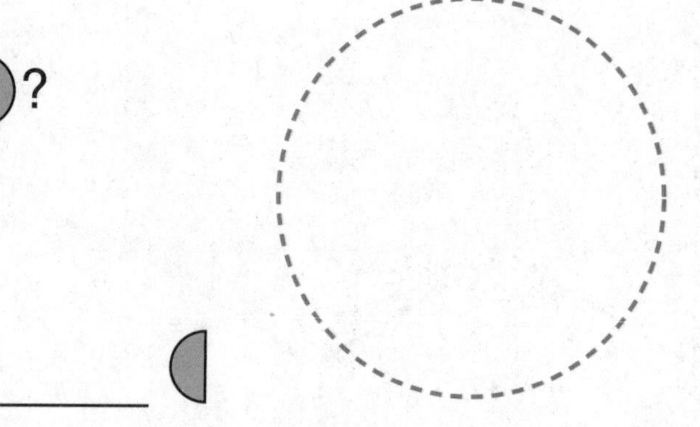

____ ◗

HOT Problem If the 4 shapes are combined, what shape do they make? Tell how you know.

Name

My Homework

Lesson 5
Composite Shapes

Homework Helper

Need help? connectED.mcgraw-hill.com

You can put shapes together to make a new shape.

Practice

Circle the two pattern blocks used to make each shape. Draw a line to show your model.

1.

2.

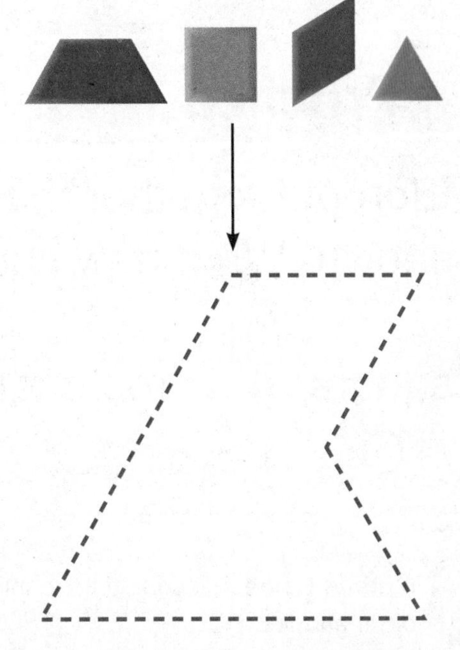

Circle the two pattern blocks used to make each shape. Draw a line to show your model.

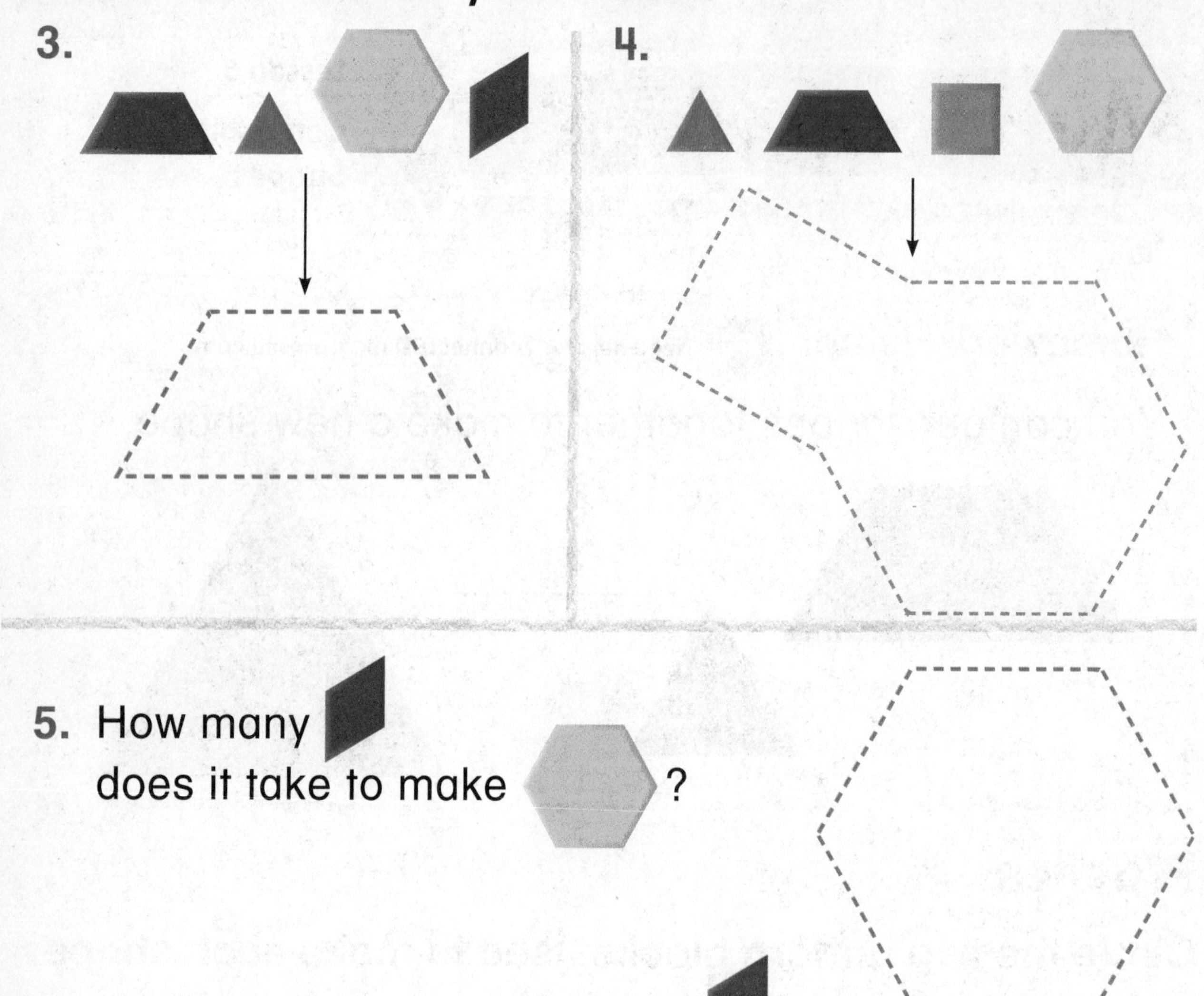

5. How many [rhombus] does it take to make [hexagon]?

______ [rhombus]

Test Practice

6. Elan put together the two shapes shown. What new shape did he make?

square ○ rectangle ○ triangle ○ trapezoid ○

Math at Home Ask your child to tell you how to compose a rectangle using two other shapes.

Name

More Composite Shapes

Lesson 6

ESSENTIAL QUESTION
How can I recognize two-dimensional shapes and equal shares?

Explore and Explain

Teacher Directions: Put two pattern blocks together. Trace the shape you made. Draw a line to show your model. Now put those pattern blocks together in a different way. Trace the new shape you made. Draw a line to show your model.

See and Show

You can move pattern blocks around to make different composite shapes.

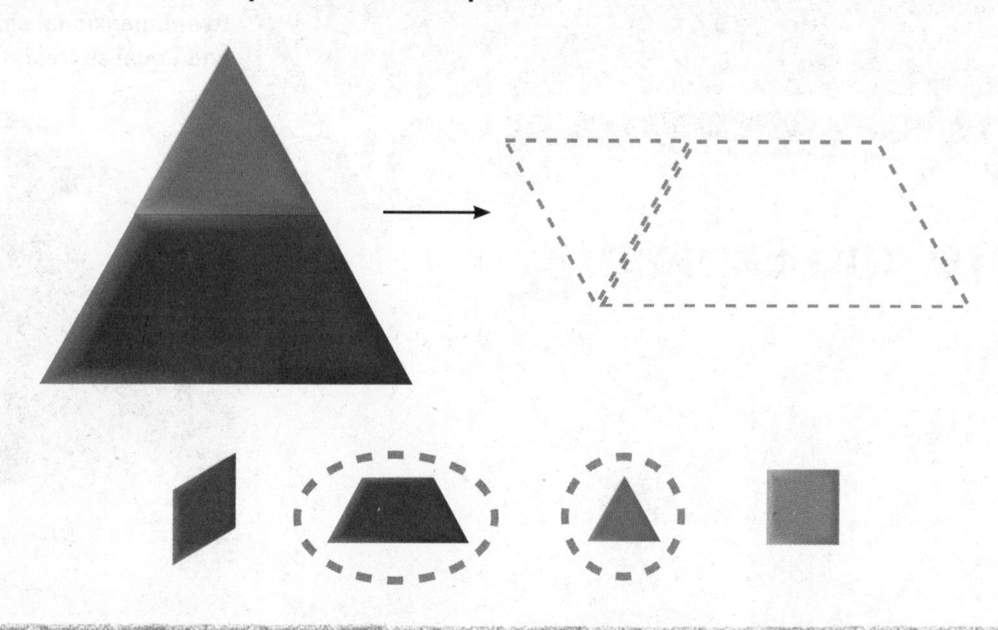

Circle the pattern blocks used to make the composite shape. Then use the same blocks to make a new shape. Draw your shape.

1.

Talk Math Describe two shapes you could put together to make a rectangle.

CCSS

Name

On My Own

Use the pattern blocks shown to make the composite shape. Then use the same blocks to make a new shape. Draw your shape.

2.

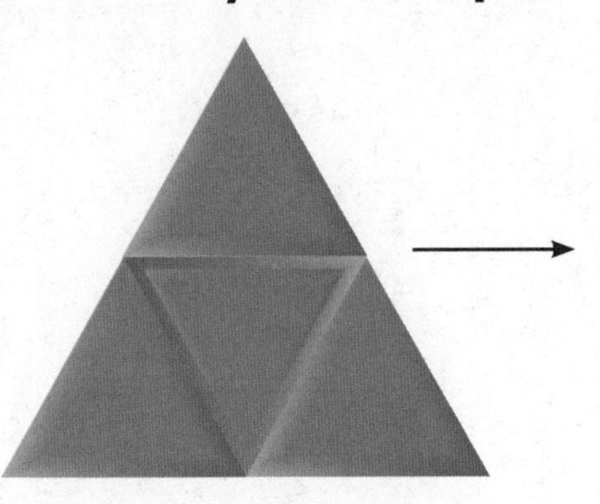

3.

4.

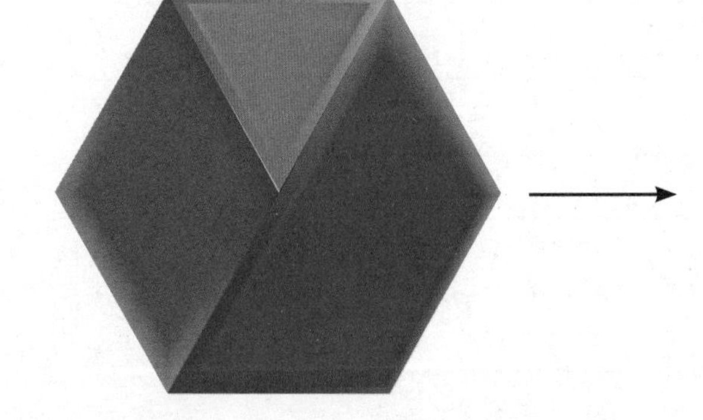

Problem Solving

5. Edgar is making different shapes using these 4 pattern blocks. Draw 1 of the shapes Edgar can make.

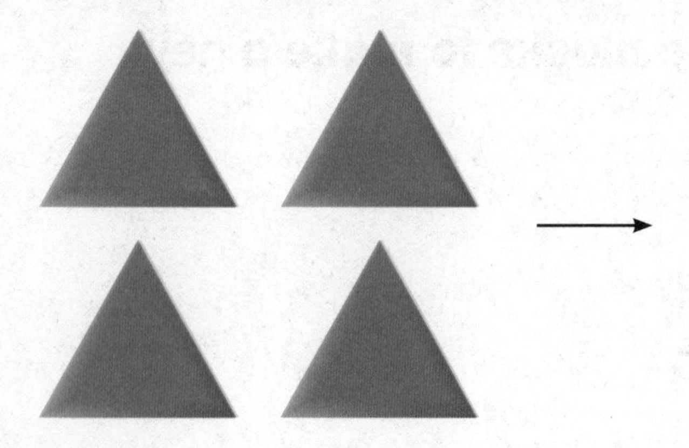

HOT Problem Alex has a hexagon. He cuts it apart into a trapezoid and 3 triangles. Draw lines to show this. Explain your answer.

Name

My Homework

Lesson 6
More Composite Shapes

Homework Helper

Need help? connectED.mcgraw-hill.com

You can move shapes around to make different shapes.

Which 3 shapes were used to make the shape above?

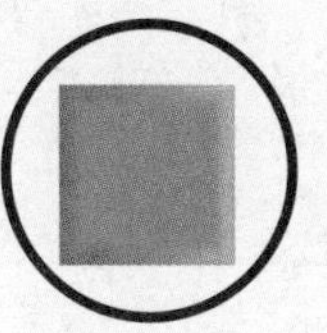

Practice

Circle the pattern blocks used to make the shape.

1.

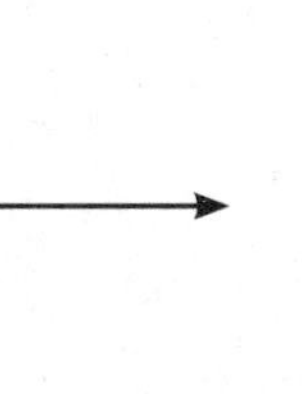

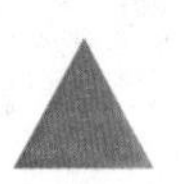

Circle the pattern blocks used to make the shape.

2.

3. Everet pushes these 2 squares together. Write the name of the new shape that Everet makes.

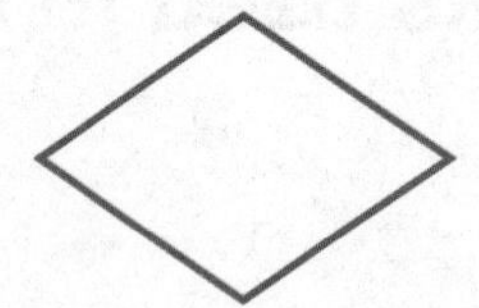

____

Test Practice

4. Sarai joined two shapes together. She made the following shape. What 2 shapes did Sarai join together?

2 triangles 

2 squares 

2 circles 

2 trapezoids 

Math at Home Cut out some triangles, squares, and rectangles from construction paper. Have your child put the shapes together to create new shapes.

Name

Problem Solving

STRATEGY: Use Logical Reasoning

Lesson 7

ESSENTIAL QUESTION
How can I recognize two-dimensional shapes and equal shares?

Ana made the shape below with 9 blocks. What 4 blocks are missing?

1-2-3 build!

1 Understand Underline what you know.
Circle what you need to find.

2 Plan How will I solve the problem?

3 Solve I will use logical reasoning.

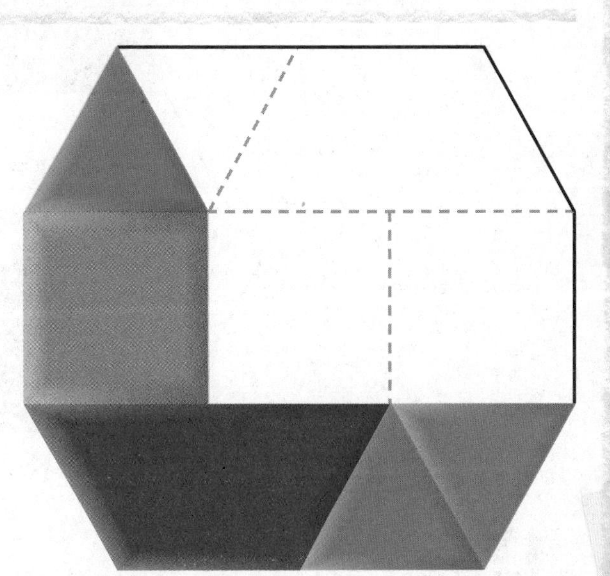

4 Check Is my answer reasonable? Explain.

Practice the Strategy

Angel has 6 pattern blocks. He makes the shape below. Two pattern blocks are missing. What two blocks are missing?

1 Understand Underline what you know.
Circle what you need to find.

2 Plan How will I solve the problem?

3 Solve I will . . .

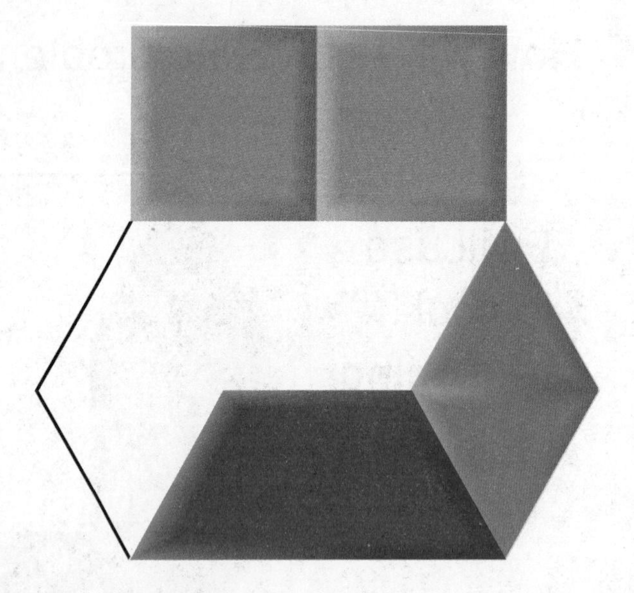

4 Check Is my answer reasonable? Explain.

Name ....

Apply the Strategy

Use pattern blocks to solve.

1. Kadence made the shape below using only trapezoids and triangles. How many blocks are missing?

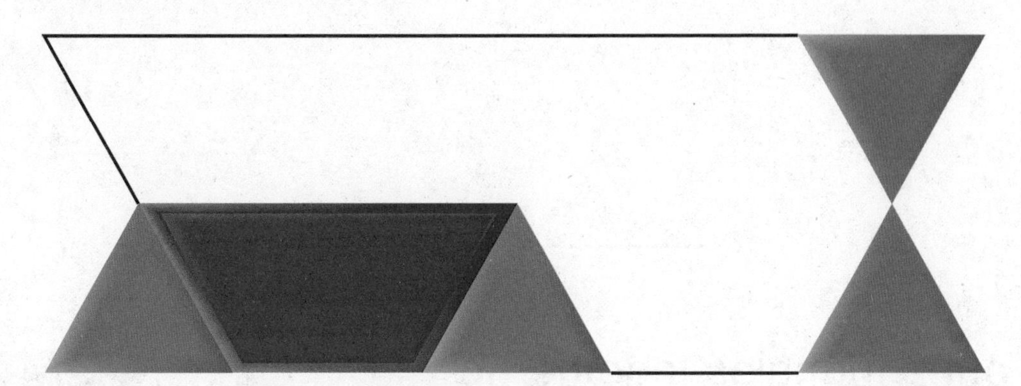

3 trapezoids and _____ triangles

2. Kendall made the shape below using triangle blocks. How many triangles are missing?

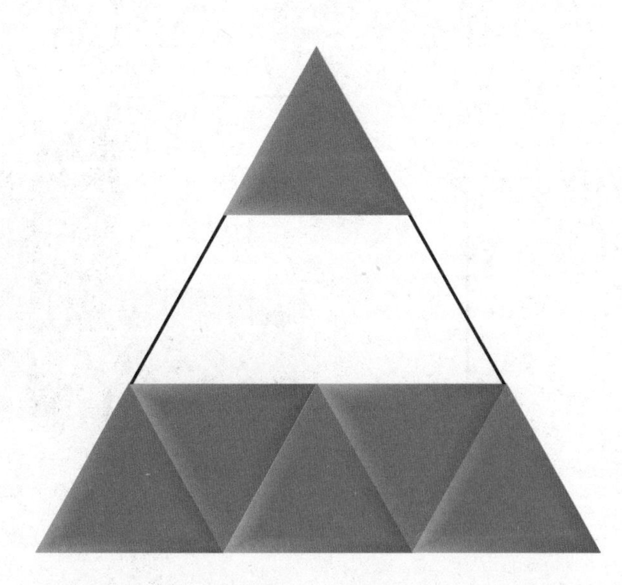

_____ triangles

Review the Strategies

Choose a strategy

- Use logical reasoning.
- Act it out.
- Draw a diagram.

3. Lynn covers this pattern block with the same 6 pattern blocks. What blocks did she use?

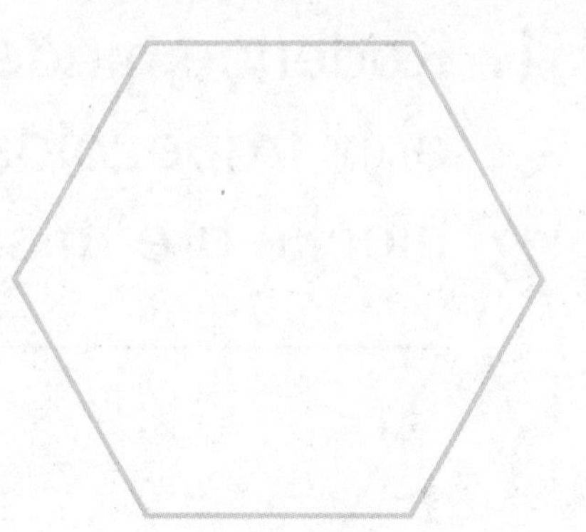

____

4. Mary put 2 triangles together to make a new shape. How many sides does the new shape have?

______ sides

5. Kevin made a shape. Kieran took 3 blocks away. Draw lines to show what 3 blocks are missing.

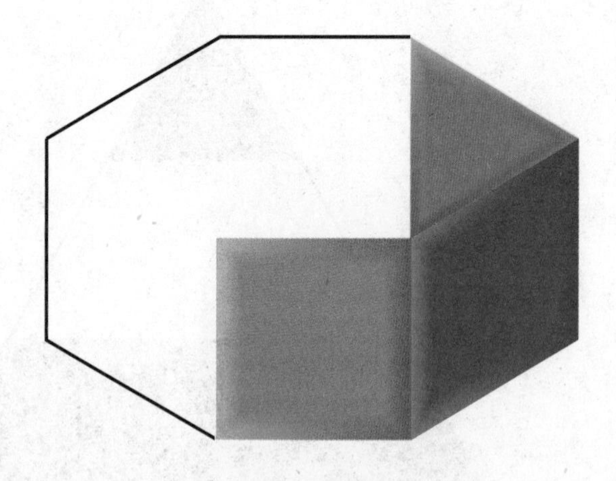

Name

My Homework

Lesson 7
Problem Solving: Use Logical Reasoning

Homework Helper

Need help? connectED.mcgraw-hill.com

Mikey made the composite shape below.
What blocks are missing?

1 Understand Underline what you know.
Circle the question.

2 Plan How will I solve the problem?

3 Solve I will use logical reasoning.

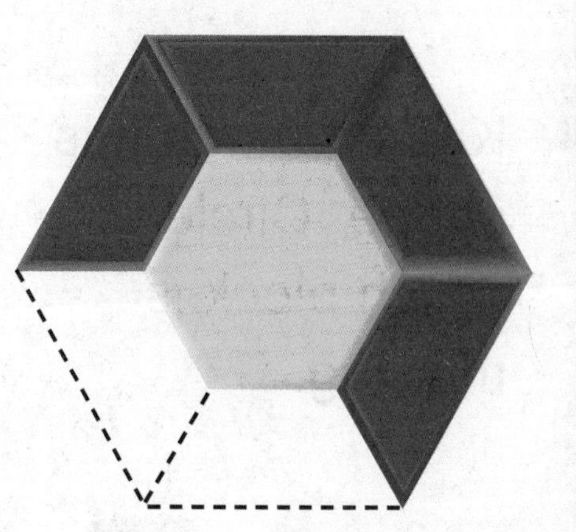

Two trapezoids are missing.

4 Check Is my answer reasonable?

Problem Solving

Underline what you know. Circle what you need to find. Use logical reasoning to solve.

1. Rashad covered the pattern block with the same three blocks. Circle which blocks he used.

2. How many trapezoids would you need to make this shape?

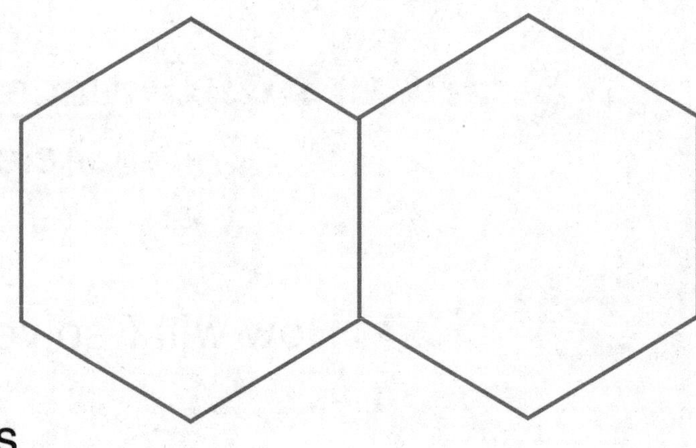

_______ trapezoids

3. Kristen made this shape. Circle which block is missing.

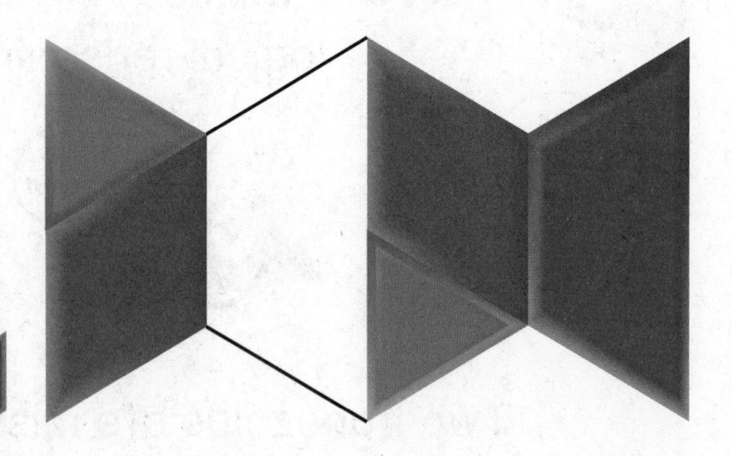

Math at Home Have your child build a shape out of construction paper squares, triangles, and trapezoids. Then, take some shapes away. Have your child figure out what shapes are missing.

Name ....

Check My Progress

Vocabulary Check

Draw lines to match.

1. side

2. vertex

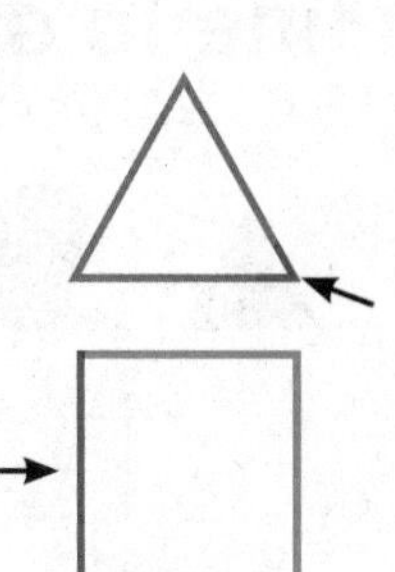

Concept Check

Write how many sides and vertices.

3.

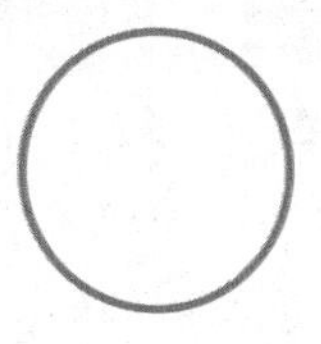

______ sides

______ vertices

4.

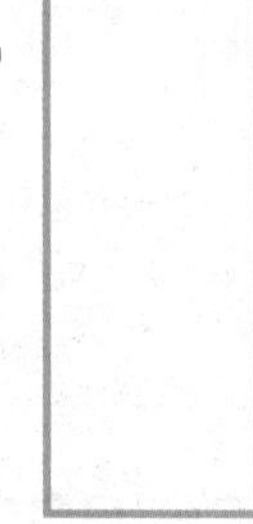

______ sides

______ vertices

5.

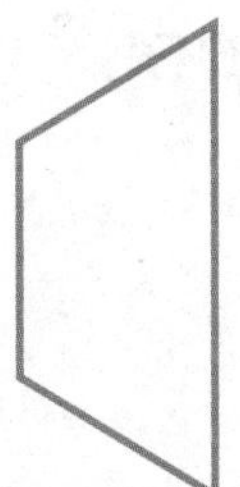

______ sides

______ vertices

6.

______ sides

______ vertices

7. Circle all of the closed shapes.

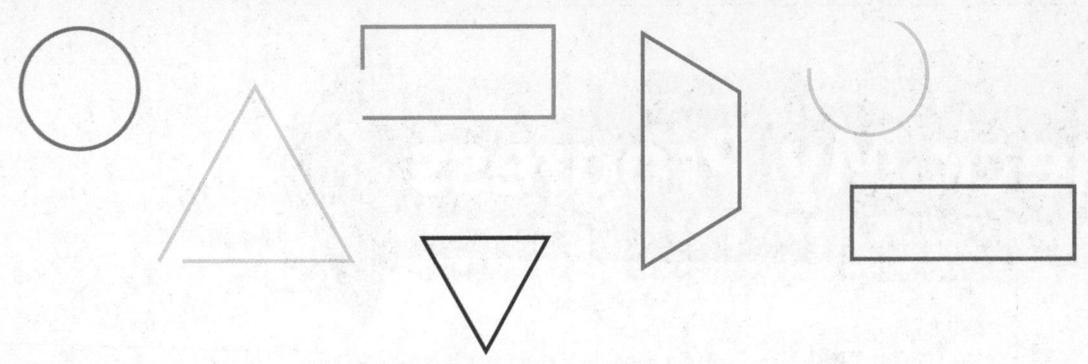

Circle the two pattern blocks used to make the shape. Draw a line to show your model.

8.

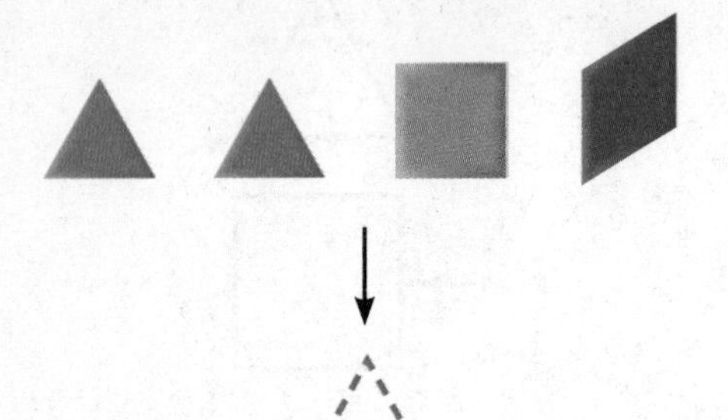

9.

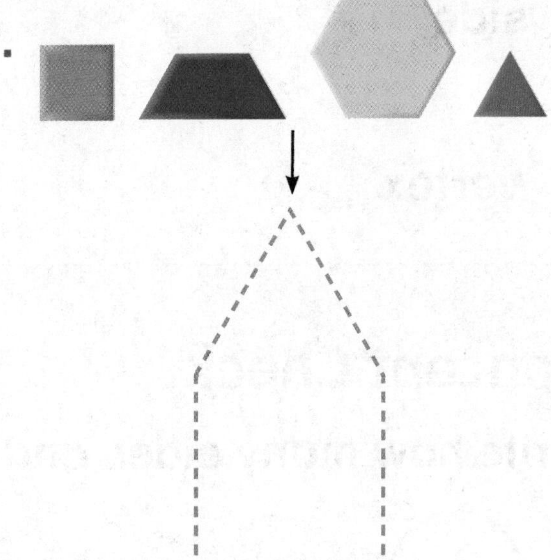

Test Practice

10. Jayne and Reece each have a shape. Jayne's shape has no vertices. Reece's shape has 4 sides that are all the same length. What are the two shapes?

rectangle, circle ○

trapezoid, square ○

square, rectangle ○

circle, square ○

Name

Geometry

1.G.3

CCSS

Equal Parts

Lesson 8

ESSENTIAL QUESTION
How can I recognize two-dimensional shapes and equal shares?

Looks equal to me!

Explore and Explain

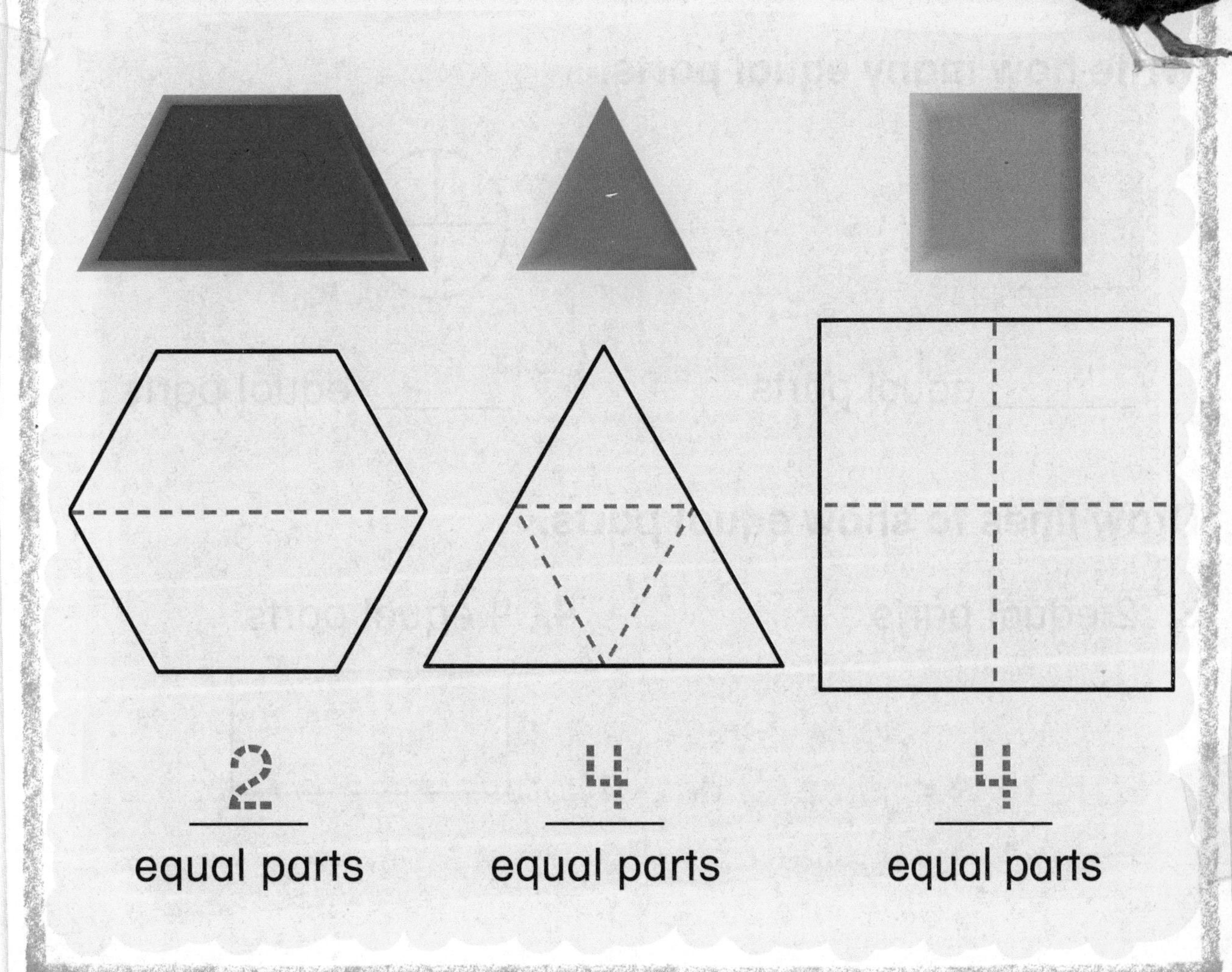

Teacher Directions: Use square, triangle, and trapezoid pattern blocks. Cover each shape with the blocks shown. Trace the lines to show your work. Tell the number of equal parts in each shape. Trace how many equal parts.

See and Show

A **whole** can be separated into **equal parts** or equal shares. Equal parts of the whole are the same size.

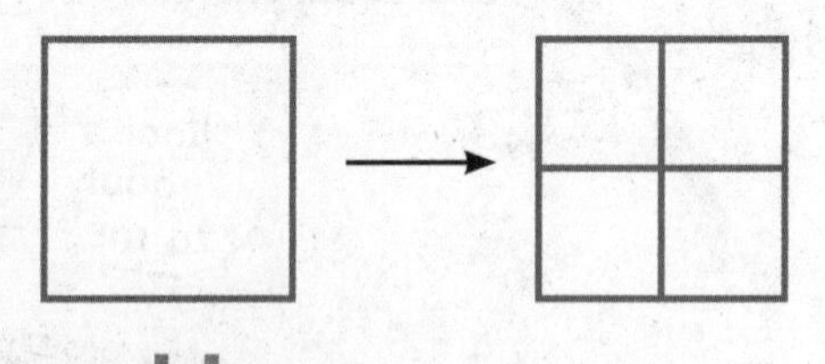

__4__ equal parts

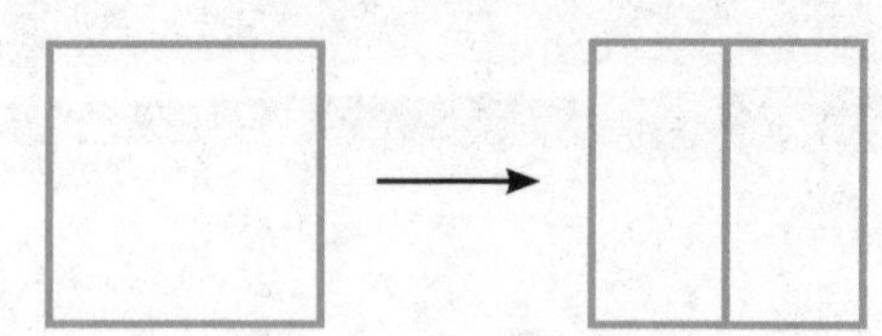

__2__ equal parts

Write how many equal parts.

1\.

______ equal parts

2\. 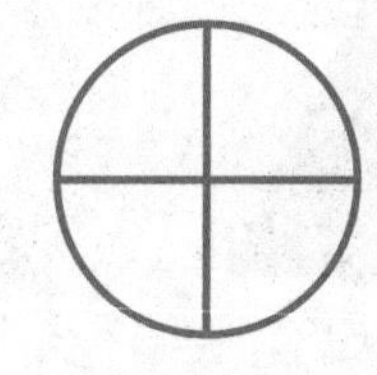

______ equal parts

Draw lines to show equal parts.

3\. 2 equal parts

4\. 4 equal parts

Talk Math How do you know when parts are equal?

Name

On My Own

Write how many equal parts.

5. 

_______ equal parts

6. 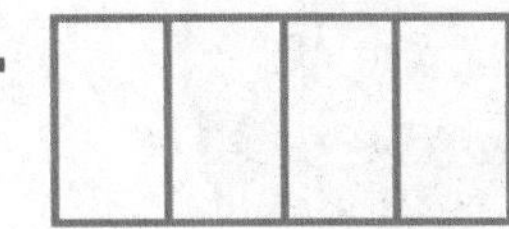

_______ equal parts

7. 

_______ equal parts

8. 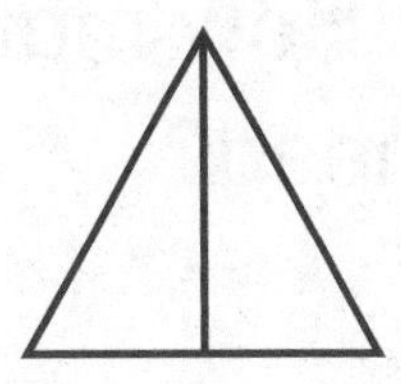

_______ equal parts

Draw lines to show equal parts.

9. 4 equal parts

10. 2 equal parts

Circle the shape that shows equal parts.

11. 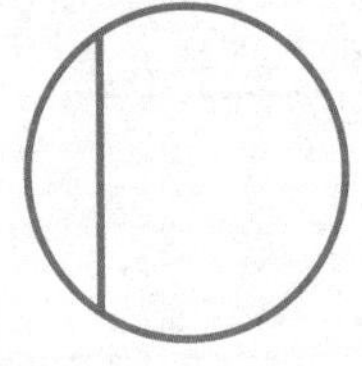  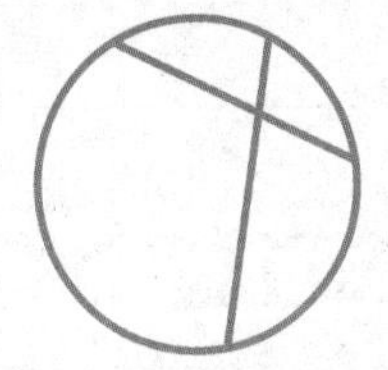

Problem Solving

12. Jasmine cuts her sandwich into two equal parts. Circle Jasmine's sandwich.

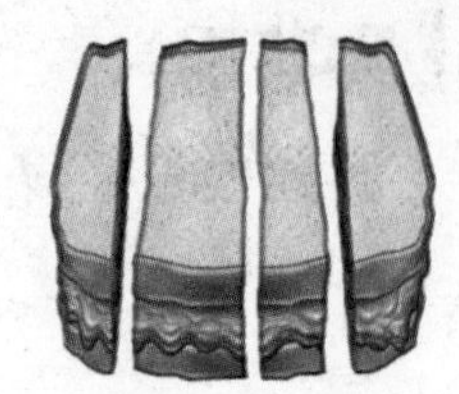
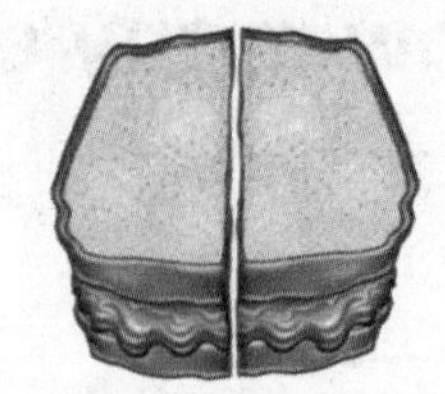

13. Damon is sharing a pie equally with 3 friends. How many equal parts does he need?

______ equal parts

HOT Problem Isabel and Katie cut this pizza to share with their two friends. Tell why they are wrong. Make it right.

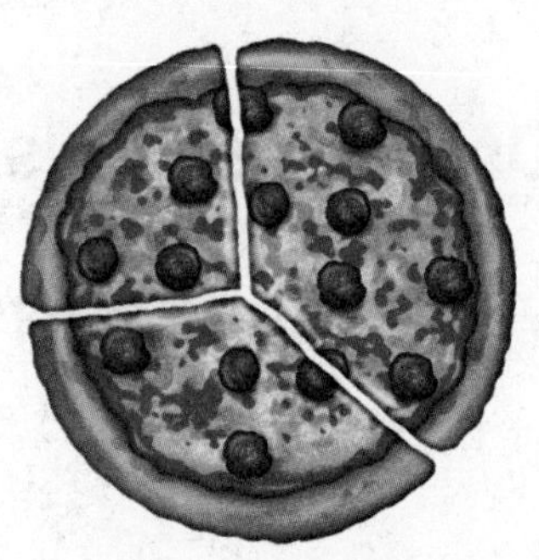

Name ________________

My Homework

Lesson 8
Equal Parts

Homework Helper

Need help? connectED.mcgraw-hill.com

A whole can be separated into equal parts.
Equal parts of the whole are the same size.

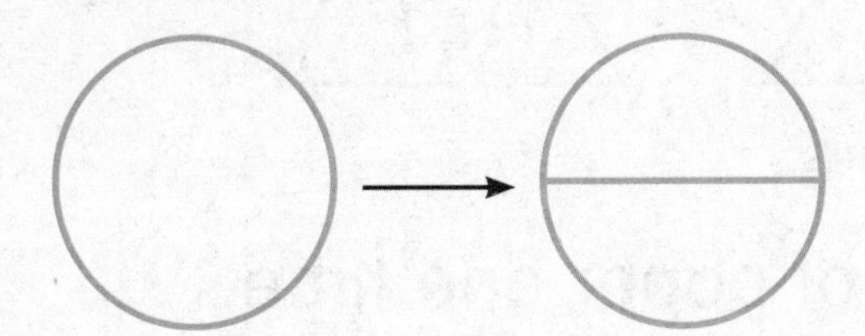

2 equal parts

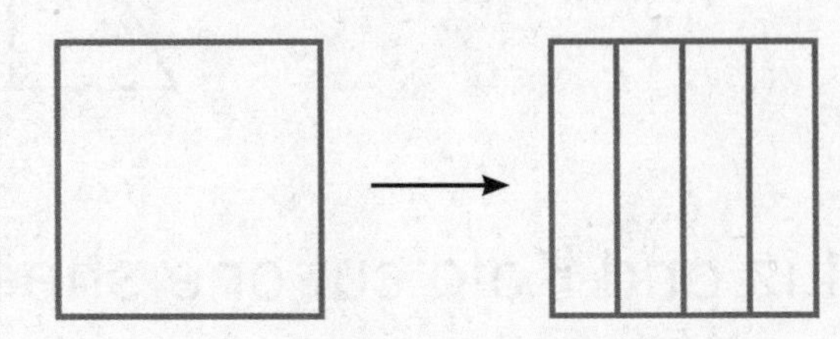

4 equal parts

Practice

Write how many equal parts.

1.

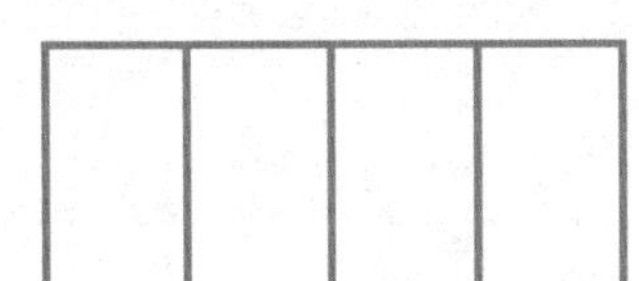

______ equal parts

2.

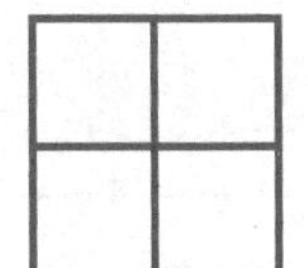

______ equal parts

3.

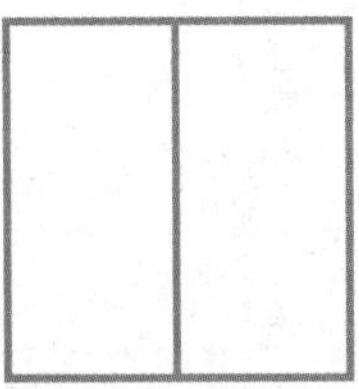

______ equal parts

4.

______ equal parts

Draw lines to show equal parts.

5. 4 equal parts

6. 2 equal parts

7. Circle the shape that shows equal parts.

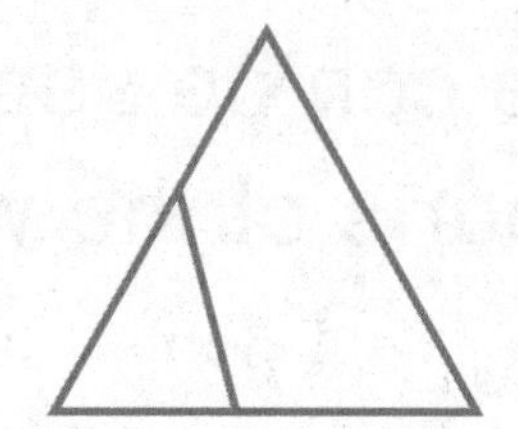

8. Liz and Koto cut one sheet of paper one time. Each gets an equal share of the whole. How many equal parts do they have?

______ equal parts

Vocabulary Check

Circle the correct answers.

9. **whole**

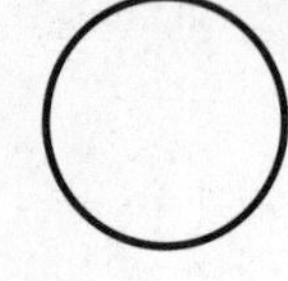

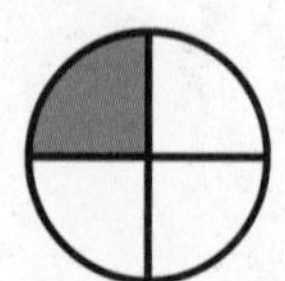

10. **equal parts**

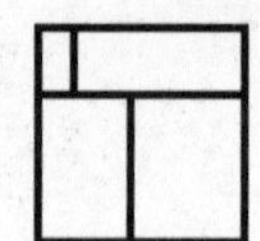

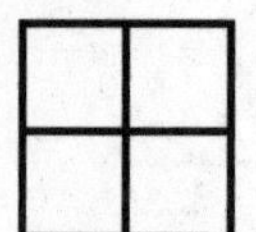

Math at Home Have your child separate a piece of toast into 2 and then 4 equal parts.

Name

Halves

Lesson 9

ESSENTIAL QUESTION
How can I recognize two-dimensional shapes and equal shares?

Explore and Explain

Draw a pretty pig pen!

Teacher Directions: Trace a square, a circle, and a rectangle attribute block to make three pig pens. Draw lines to separate each shape into two equal parts. Shade each part a different color.

See and Show

A whole that is separated into 2 equal parts is separated into **halves**.

Each part is a half of the whole.

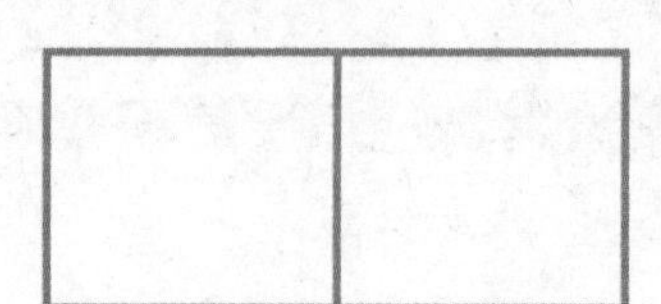

__2__ equal parts, or __2__ halves

Write how many equal parts make up the whole.

1. 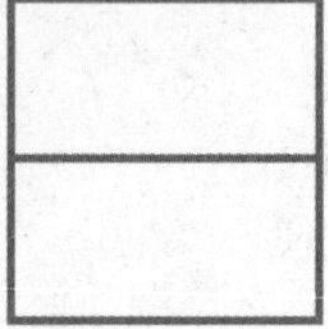

_____ equal parts

2. 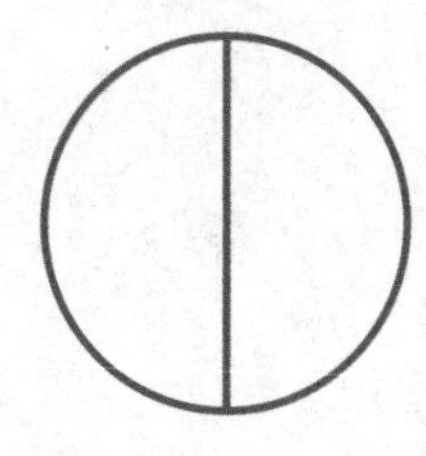

_____ equal parts

Draw lines to show two equal parts. Write how many halves.

3. 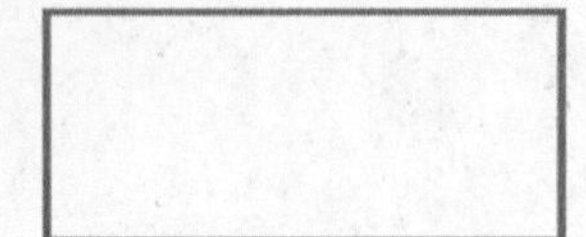

_____ halves

4. 

_____ halves

Talk Math How many halves make up a whole?

Name

CCSS

On My Own

Write how many equal parts make up the whole.

5. 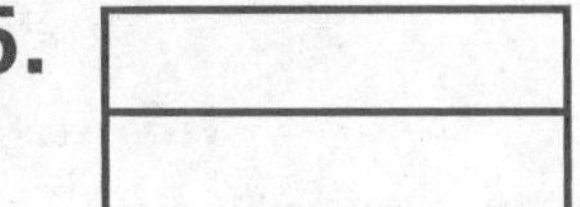

_______ equal parts

6. 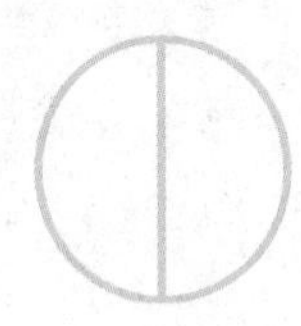

_______ equal parts

A half of each shape is missing. Draw the missing half.

7. 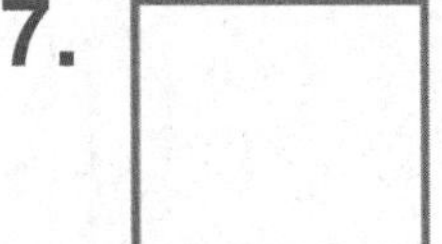

8. 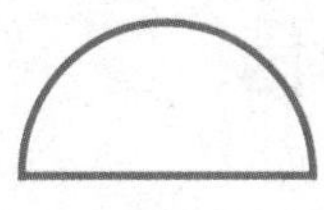

9. 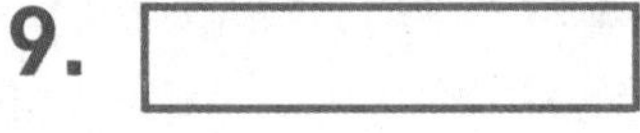

10. 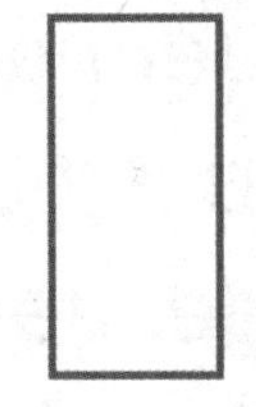

Color half of each shape. Write how many parts are shaded.

11. 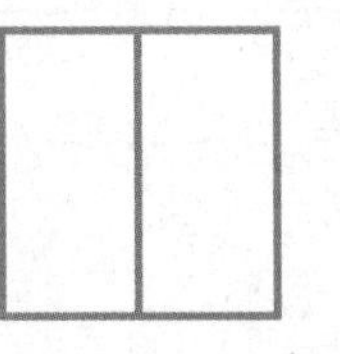

_______ of _______ parts

12. 

_______ of _______ parts

Problem Solving

13. Joey has a hot dog. He cuts it in half. How many equal parts is the hot dog cut into?

_______ equal parts

14. Tina draws this square. Help Tina by drawing a line to show 2 equal parts.

HOT Problem Jenna is eating half of an orange. Samantha is eating the other half of the same orange. Jenna says she has less than Samantha. Can she be right?

Name ....

My Homework

Lesson 9
Halves

Homework Helper

eHelp  Need help? connectED.mcgraw-hill.com

A shape that is divided into 2 equal parts is divided into halves.

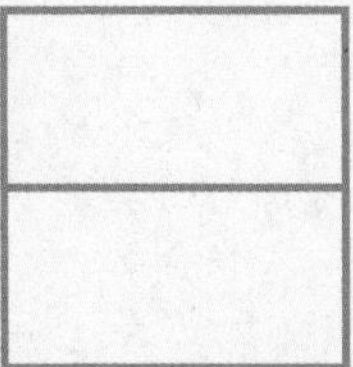

2 equal parts, or 2 halves

Practice

Write how many equal parts make up the whole.

1. 

_____ equal parts

2. 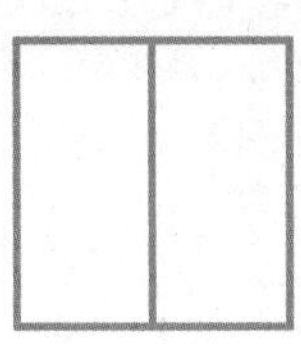

_____ equal parts

Draw lines to show two equal parts. Write how many halves.

3. 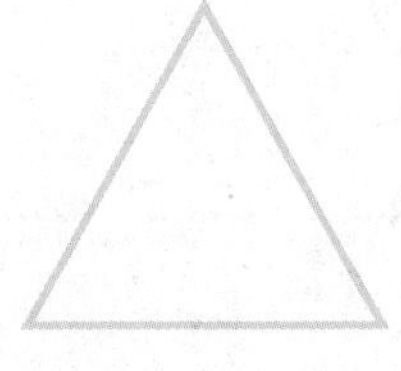

_____ halves

4. 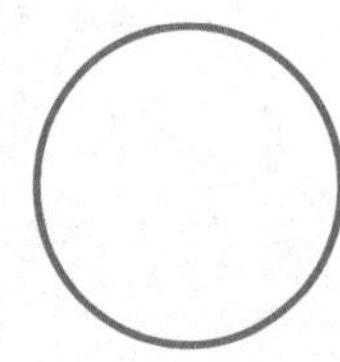

_____ halves

A half of each shape is missing. Draw the missing half.

5. 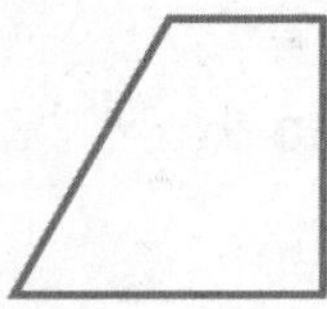

6. 

Color half of each shape. Write how many parts are shaded.

7. 

_____ of _____ parts

8. 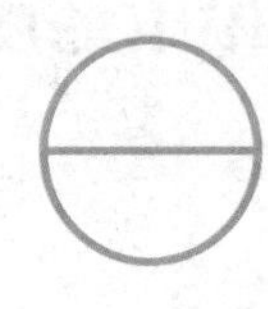

_____ of _____ parts

9. Two friends share an apple equally. How many equal parts are there?

_____ equal parts

Vocabulary Check

10. Circle the shape that shows **halves**.

 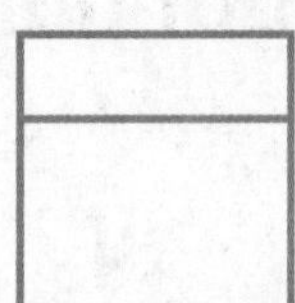 

Math at Home Ask your child to use yarn to divide the kitchen table into 2 equal parts. Have him or her describe the 2 parts in different ways, such as equal parts, 1 of 2 parts, or half of.

Name

Quarters and Fourths

Lesson 10

ESSENTIAL QUESTION
How can I recognize two-dimensional shapes and equal shares?

Psst! You'll need this!

Explore and Explain

Teacher Directions: Separate the rectangle into two equal parts. Then separate it into four equal parts. Shade each part a different color. Now separate the circle into four equal parts. Shade each part a different color.

See and Show

A whole that is separated into 4 equal parts is separated into **fourths** or quarters.

Each part is a fourth of, or a quarter of, the whole.

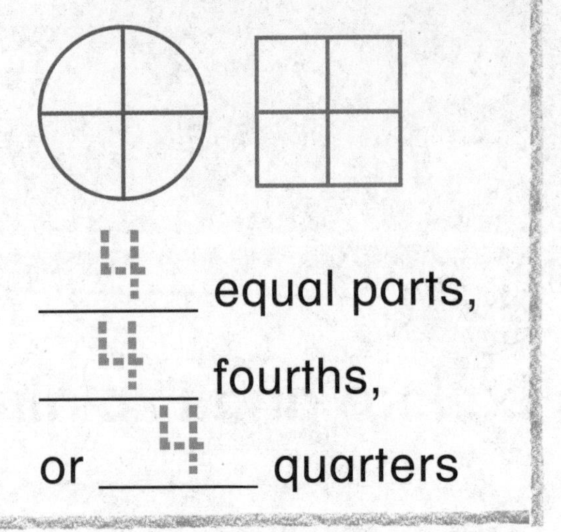

___4___ equal parts,
___4___ fourths,
or ___4___ quarters

Write how many equal parts make up the whole.

1. 

   ______ equal parts

2. 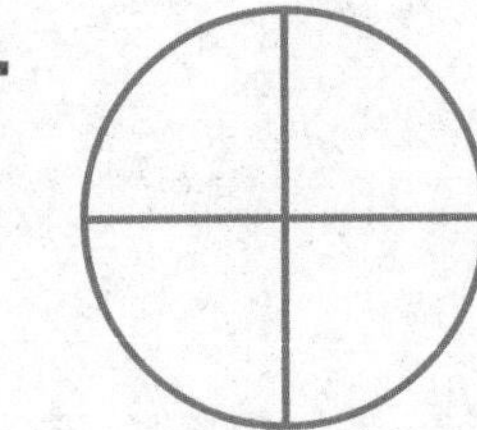

   ______ equal parts

Draw lines to show 4 equal parts. Write how many fourths.

3. 

   ______ fourths

4. 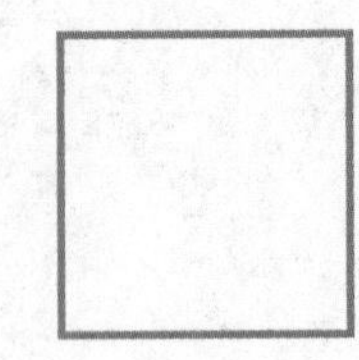

   ______ fourths

Talk Math How are halves and fourths different?

Name ______________________

On My Own

A quarter of each shape is missing. Draw the missing quarter.

5. 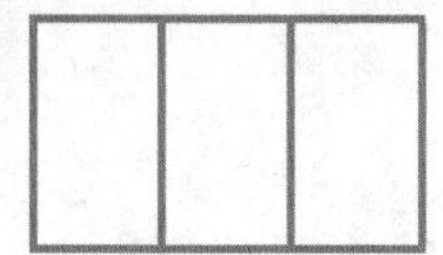

6. 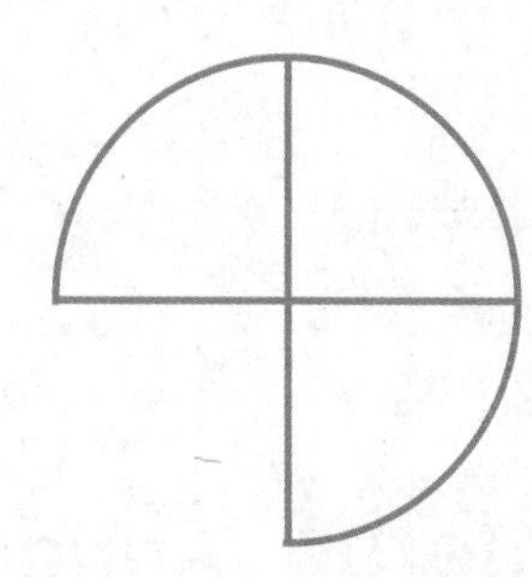

Color a fourth of each shape. Write how many parts are shaded.

7. 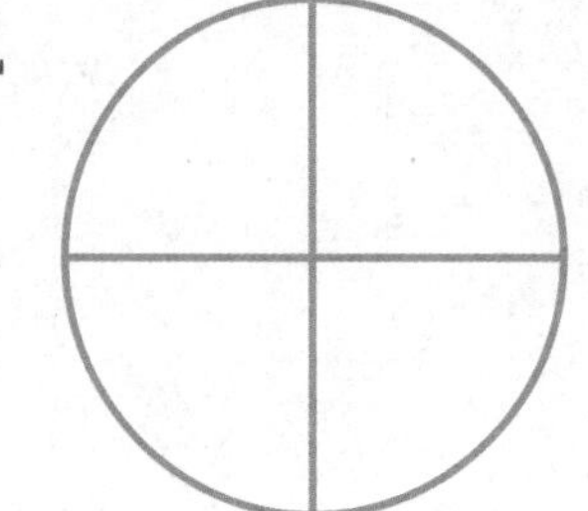

_______ of _______ parts

8. 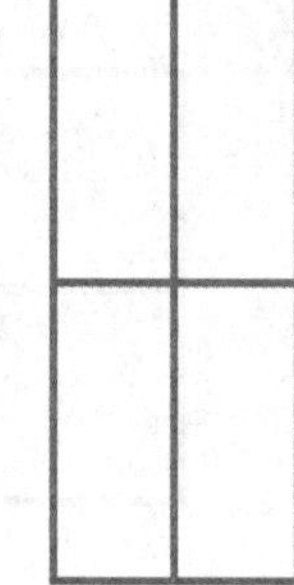

_______ of _______ parts

Color one equal part. Fill in the numbers.

9. 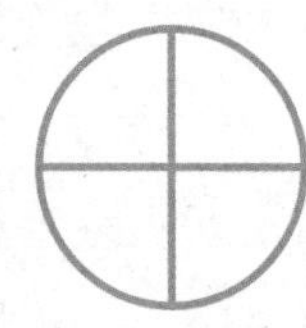

_______ of _______ parts

10. 

_______ of _______ parts

Problem Solving

11. Heidi has a cookie. She and 3 friends want to share the cookie. How many equal parts does Heidi need?

_____ equal parts

HOT Problem Adrianne draws a line on a rectangle to show equal parts. She then draws another line to show more equal parts. Describe what happens to the sizes of the parts.

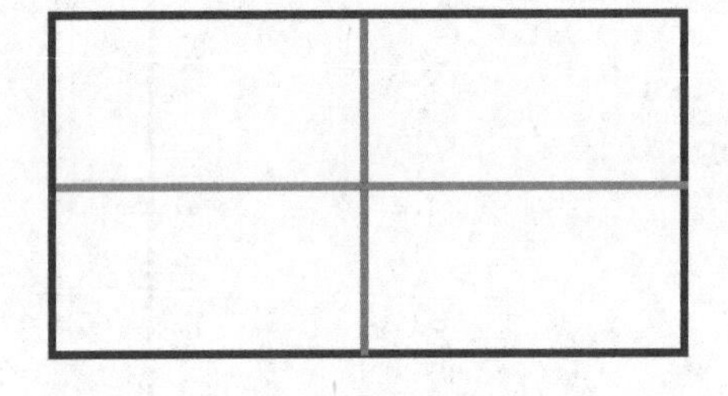

Name

My Homework

Lesson 10
Quarters and Fourths

Homework Helper

Need help? connectED.mcgraw-hill.com

A shape that is divided into 4 equal parts is divided into fourths or quarters.

4 equal parts,
4 fourths, or 4 quarters

Practice

Write how many equal parts make up the whole.

1. 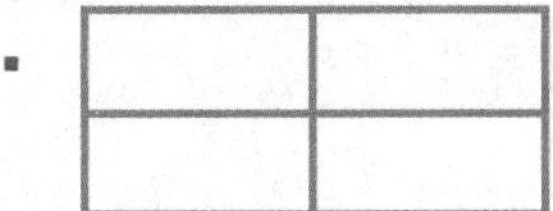

_______ equal parts

2. 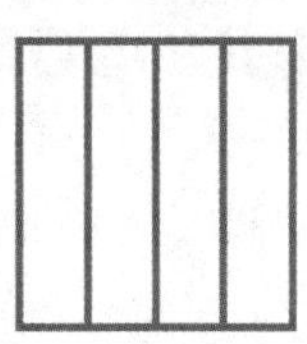

_______ equal parts

Draw lines to show 4 equal parts. Write how many fourths.

3. 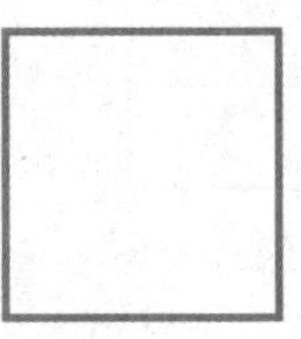

_______ fourths

4. 

_______ fourths

A quarter of each shape is missing. Draw the missing quarter.

5. 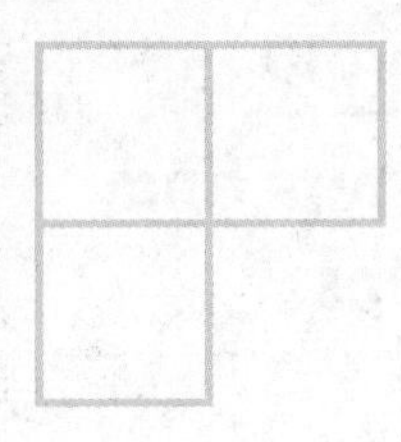

6. 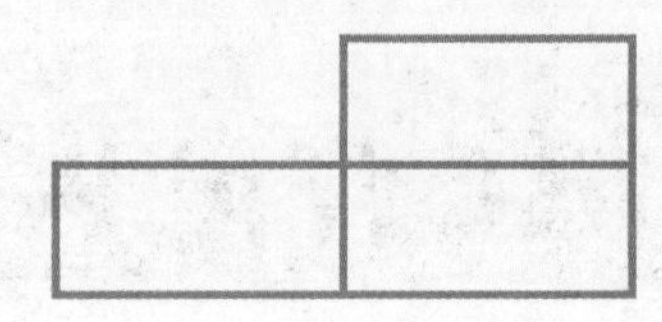

Color a fourth of each shape. Write how many parts are shaded.

7. 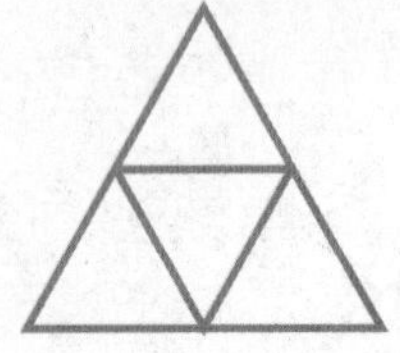

______ of ______ parts

8. 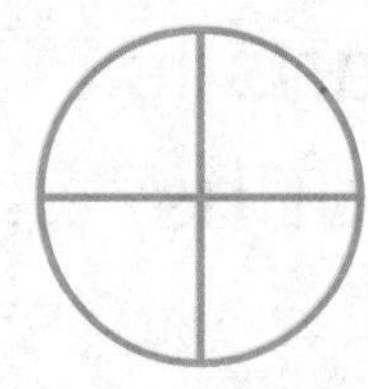

______ of ______ parts

9. Sam and three friends shared a sandwich equally. How many equal parts were there?

______ equal parts

Vocabulary Check

10. Circle the shape that shows **fourths**.

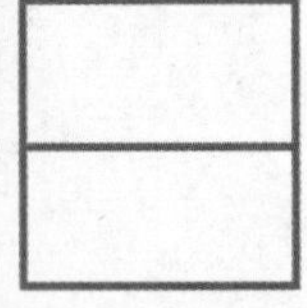

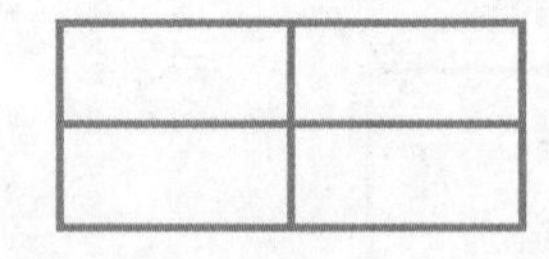

Math at Home Cut a large square, a large rectangle, and a large circle from a piece of paper. Have your child show you fourths by folding the shapes into 4 equal parts.

Name ....

My Review

Chapter 9
Two-Dimensional Shapes and Equal Shares

Vocabulary Check

Complete the sentences.

fourths **halves**

two-dimensional shape **whole**

1. Two equal parts of a whole are called ____________________.

2. A ____________________ is a flat shape, such as a circle, a triangle, or a square.

3. Four equal parts of a whole are called ____________________.

4. The entire amount or all of the parts is called the ____________________.

Concept Check

Write how many sides and vertices.

5. ______ sides

______ vertices

6. ______ sides

______ vertices

Circle the two pattern blocks used to make the shape. Draw a line to show your model.

7.

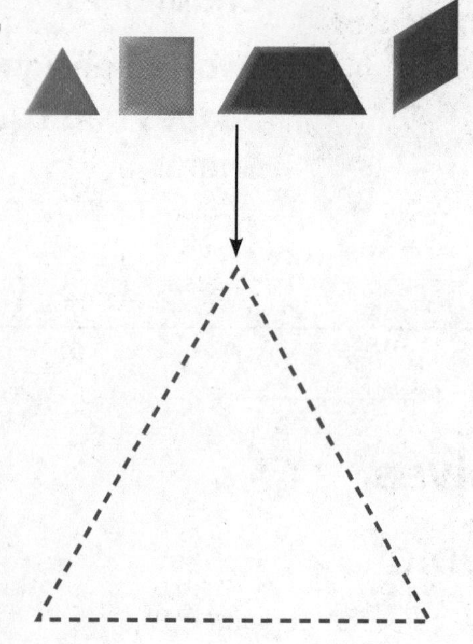

8.

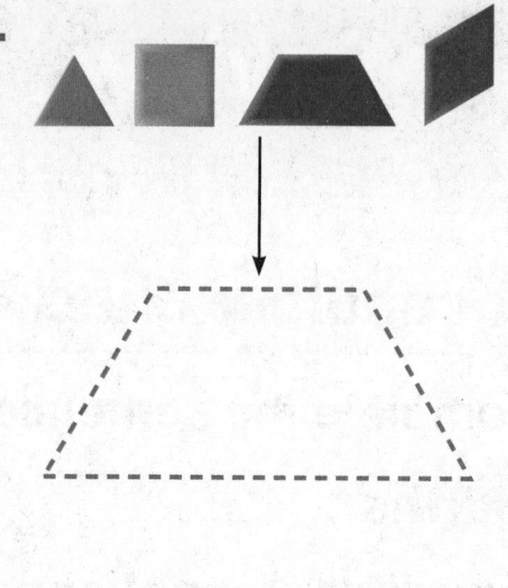

Write how many equal parts make up the whole.

9.

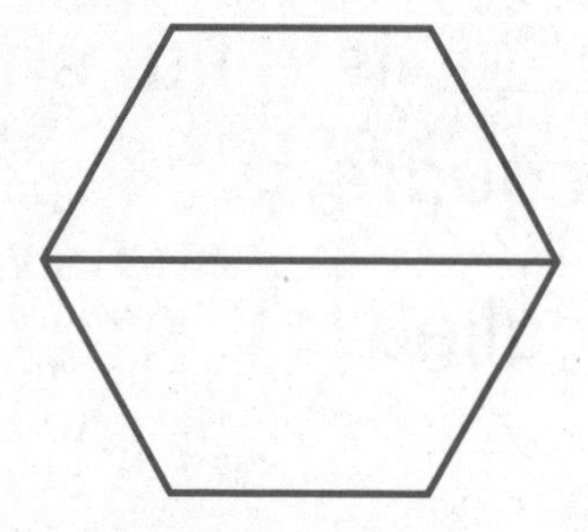

_____ equal parts

10.

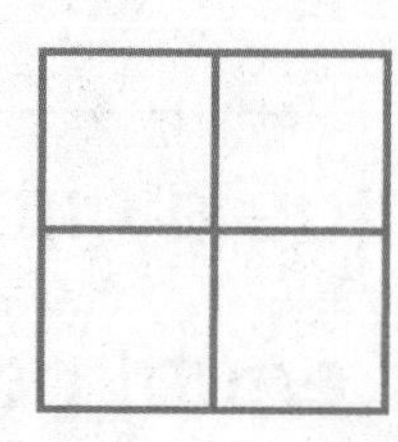

_____ equal parts

Write how many parts are shaded.

11.

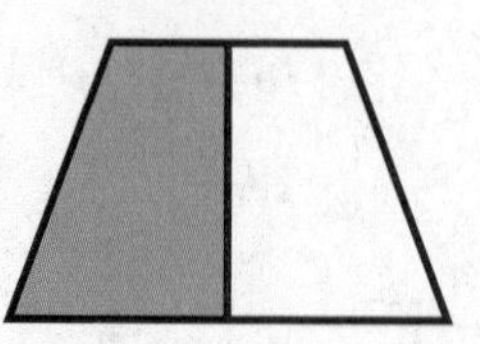

_____ of _____ parts

12.

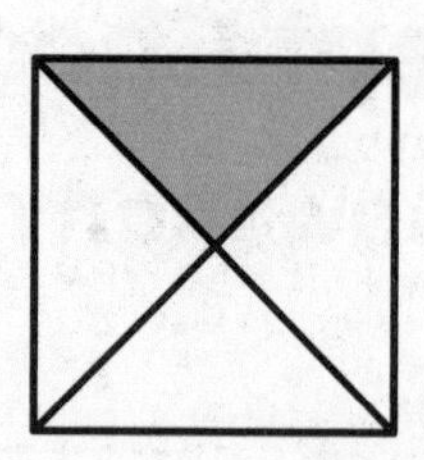

_____ of _____ parts

Name

Problem Solving

13. I am a two-dimensional shape that has more than two but less than 4 sides. What shape am I?

____

14. Amos and Elton want a piece of chicken pot pie. Each of them want an equal part of it. Draw a line to show how they should cut the pot pie.

Test Practice

15. Jill and Carlos each have a shape. Both shapes have 4 sides and 4 vertices. Jill's shape has 4 sides all the same length. Carlos's shape has exactly 1 pair of sides that are the same length. What are the two shapes?

- square and triangle ○
- square and trapezoid ○
- triangle and rectangle ○
- circle and square ○

Reflect

Chapter 9
Answering the Essential Question

ESSENTIAL QUESTION

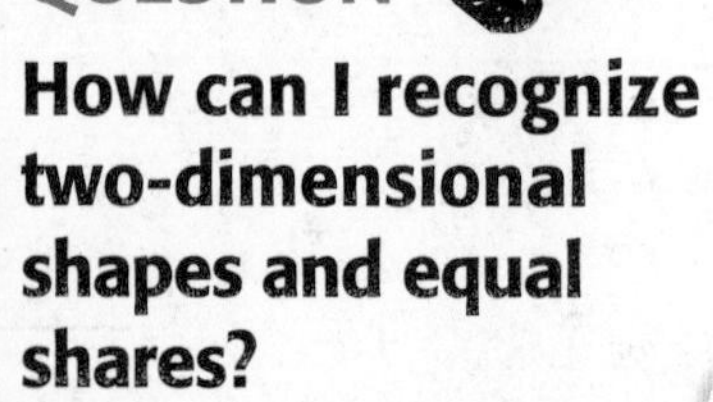

How can I recognize two-dimensional shapes and equal shares?

Circle the shapes with more than 3 vertices.

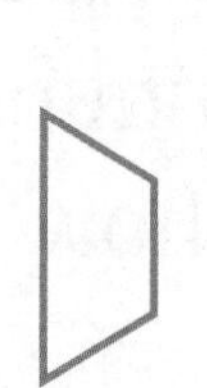

Circle the pattern blocks used to make the composite shape.

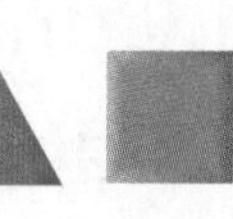

Draw lines to show equal parts.

2 equal parts

4 equal parts

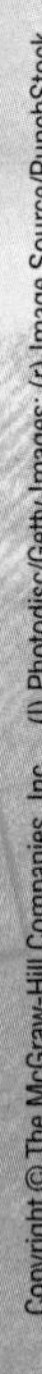

Three-Dimensional Shapes

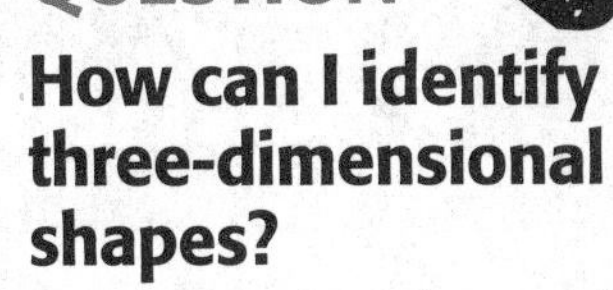

ESSENTIAL QUESTION

How can I identify three-dimensional shapes?

Our Kitchen Adventures!

Watch a video!

My Common Core State Standards

Geometry

1.G.1 Distinguish between defining attributes (e.g., triangles are closed and three-sided) versus non-defining attributes (e.g., color, orientation, overall size); build and draw shapes to possess defining attributes.

1.G.2 Compose two-dimensional shapes (rectangles, squares, trapezoids, triangles, half-circles, and quarter-circles) or three-dimensional shapes (cubes, right rectangular prisms, right circular cones, and right circular cylinders) to create a composite shape, and compose new shapes from the composite shape.

Standards for Mathematical PRACTICE

1. Make sense of problems and persevere in solving them.
2. Reason abstractly and quantitatively.
3. Construct viable arguments and critique the reasoning of others.
4. Model with mathematics.
5. Use appropriate tools strategically.
6. Attend to precision.
7. Look for and make use of structure.
8. Look for and express regularity in repeated reasoning.

= focused on in this chapter

Name

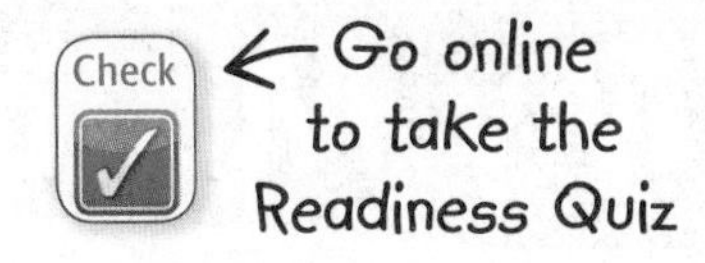

Draw an X on the object that is a different shape.

1. 

2. 

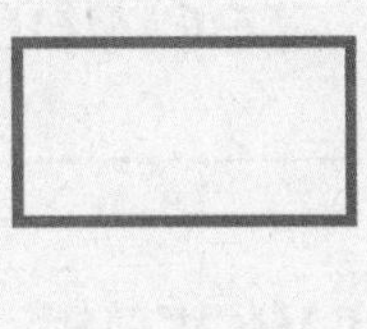

Draw lines to match the objects that are the same shape.

3. 

4. 

5. 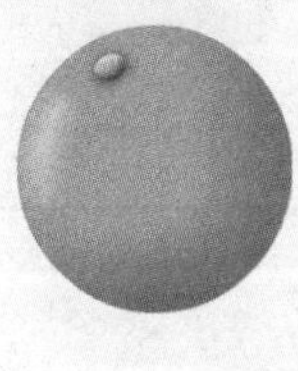

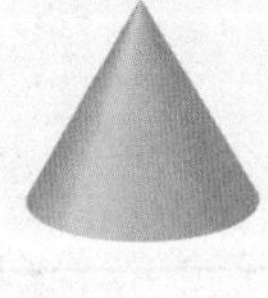

6. Miley was using this object to play a game. Circle the name of the shape.

cylinder cone cube

How Did I Do?

Shade the boxes to show the problems you answered correctly.

1	2	3	4	5	6

Name

My Math Words

Review Vocabulary

circle rectangle square

Use the review words. Write the name of each shape shown in color.

Two-Dimensional Shape Hunt

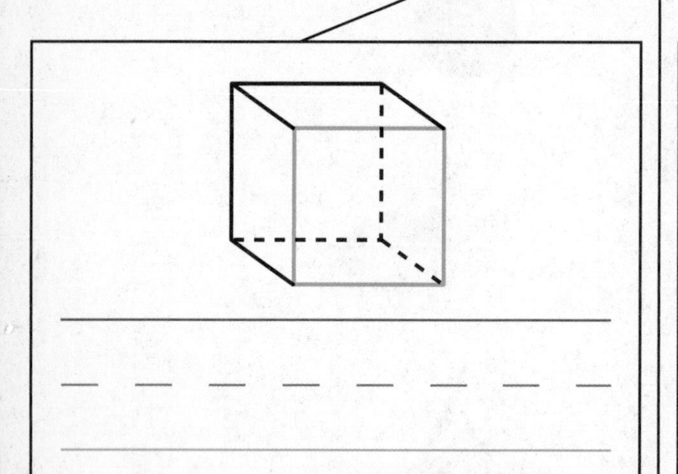

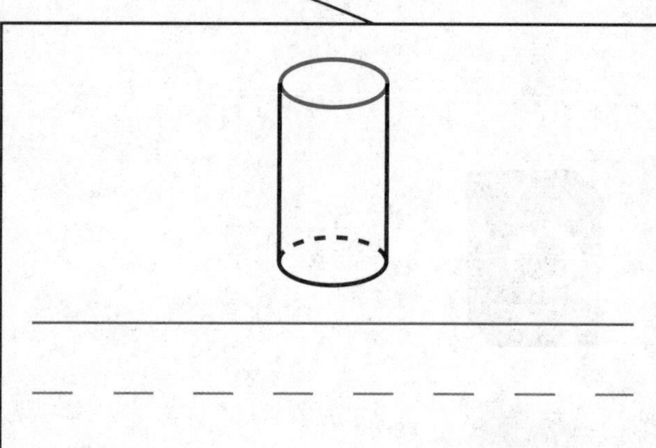

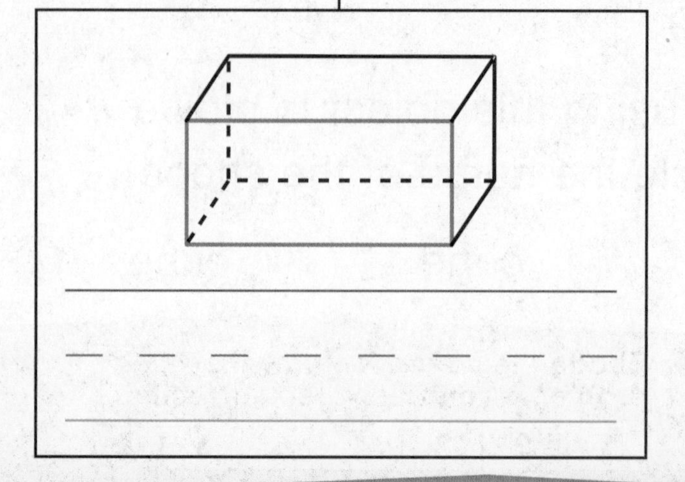

My Vocabulary Cards

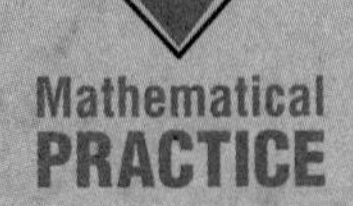

Lesson 10-2

cone

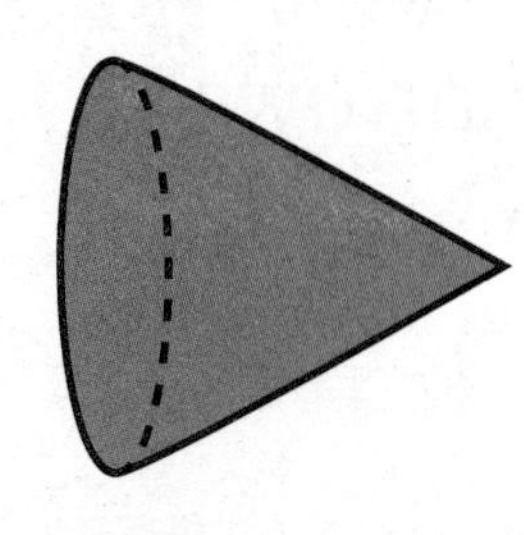

Lesson 10-1

cube

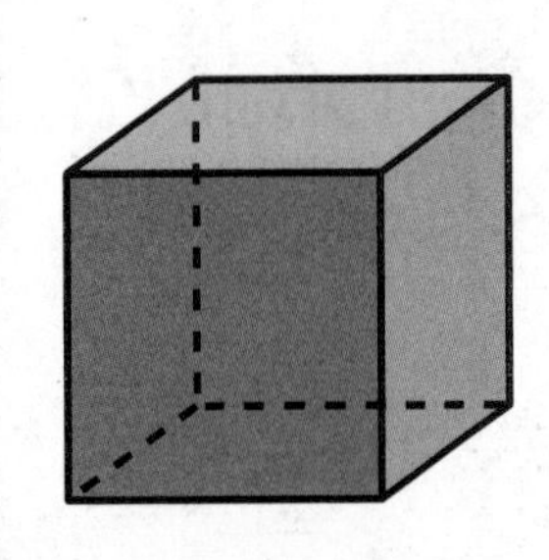

Lesson 10-2

cylinder

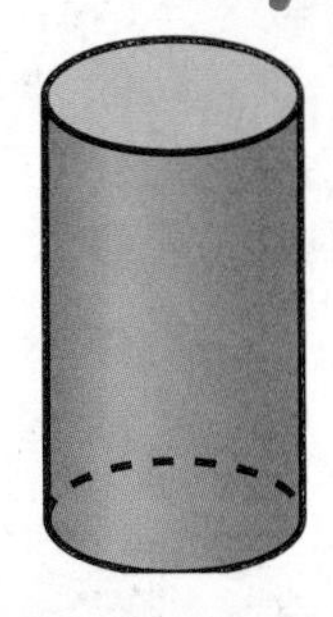

Lesson 10-1

face

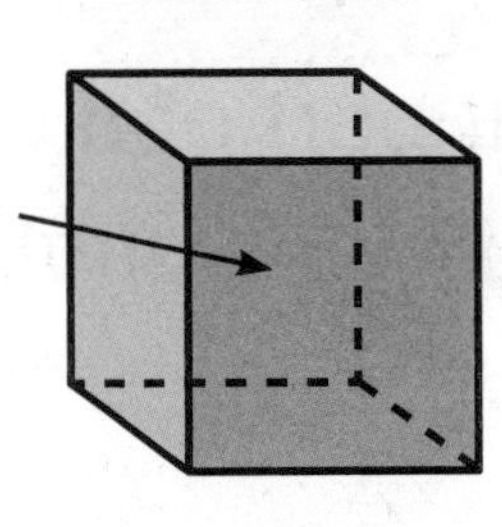

6 faces

Lesson 10-1

rectangular prism

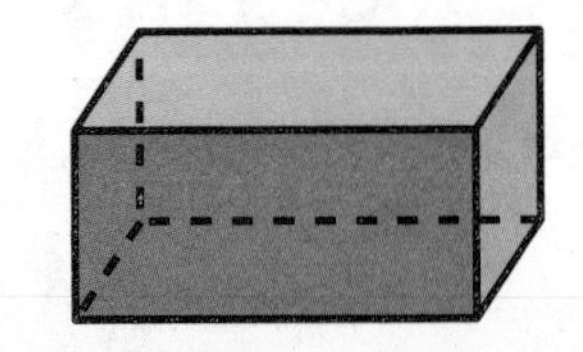

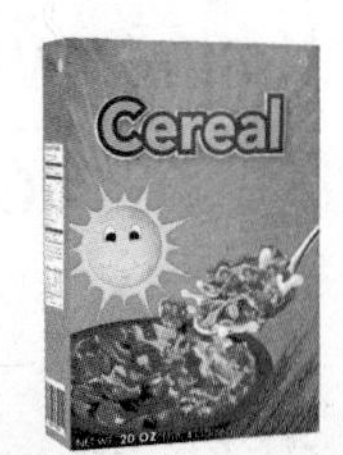

Lesson 10-1

three-dimensional shape

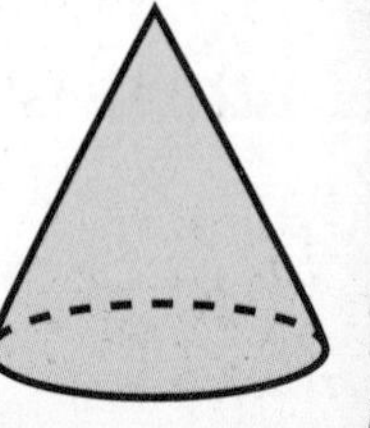

Teacher Directions:
Ideas for Use

- Tell students to create riddles for each word. Ask them to work with a friend to guess the word for each riddle.
- Have students sort the words by the number of letters in each word.

A three-dimensional shape with 6 square faces.

A three-dimensional shape that narrows to a point from a circular face.

The flat part of a three-dimensional shape.

A three-dimensional shape that is shaped like a can.

A solid shape. A shape that is not flat.

A three-dimensional shape with 6 faces that are rectangles.

My Foldable

FOLDABLES® Follow the steps on the back to make your Foldable.

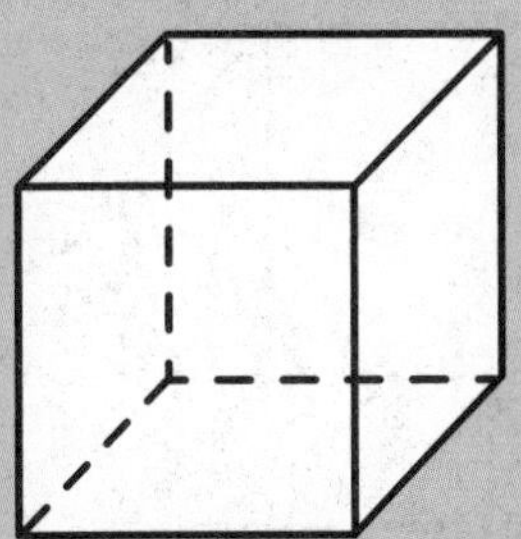

_____ faces

_____ vertices

rectangular prism

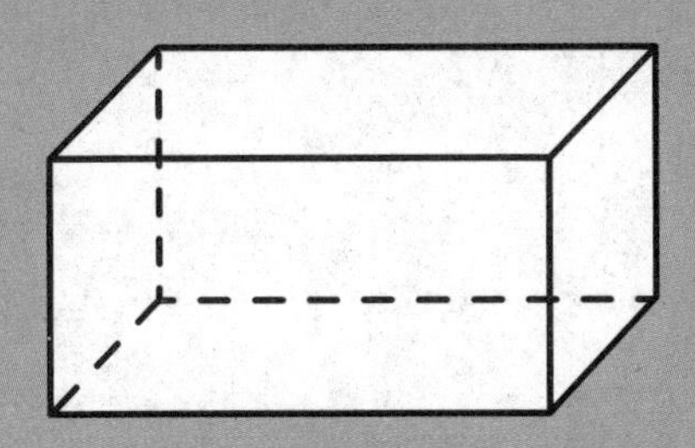

_____ faces

_____ vertices

cone

_____ face

_____ vertex

cylinder

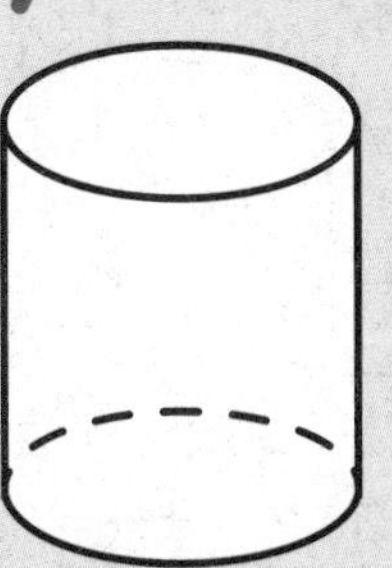

_____ faces

_____ vertices

FOLDABLES®
Study Organizer

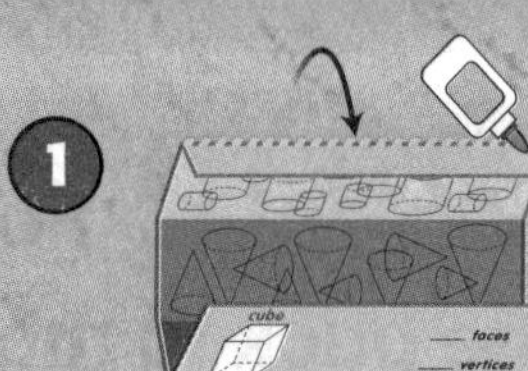
1

2

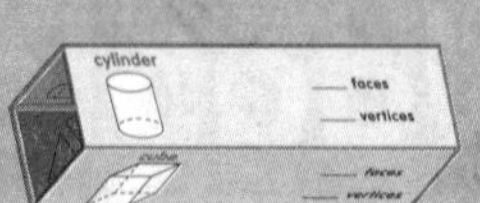

Name ______________________

Cubes and Prisms

Lesson 1

ESSENTIAL QUESTION
How can I identify three-dimensional shapes?

What shapes are we?

Explore and Explain

Teacher Directions: Use and . Compare and describe the shapes. Trace one face of each shape. Tell a classmate which three-dimensional shape belongs to each face you traced.

See and Show

Three-dimensional shapes are solid shapes. Cubes and rectangular prisms have **faces** and vertices (vertex).

cube

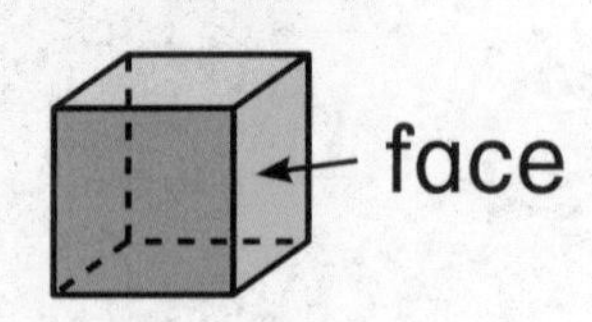

rectangular prism

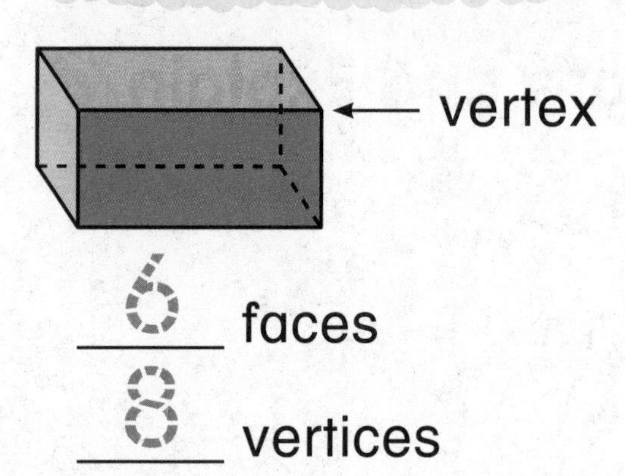

6 faces
8 vertices

6 faces
8 vertices

Identify each shape. Circle the name.
Write the number of faces and vertices.

1. 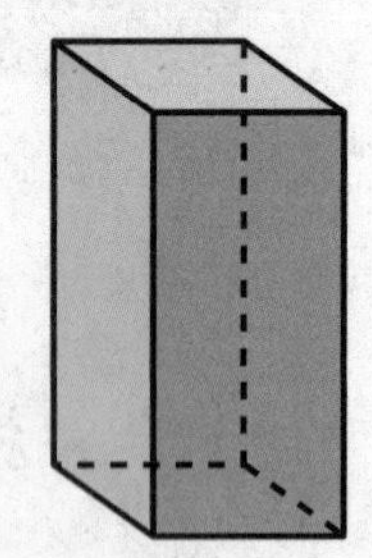

cube   rectangular prism

_____ faces   _____ vertices

2. 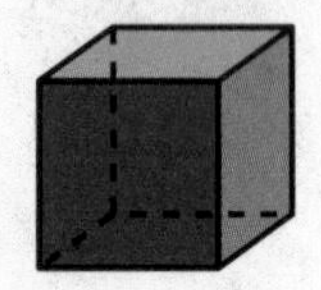

cube   rectangular prism

_____ faces   _____ vertices

Talk Math How are a rectangular prism and a cube alike?

Name ....

On My Own

Identify each shape. Circle the name.
Write the number of faces and vertices.

3. 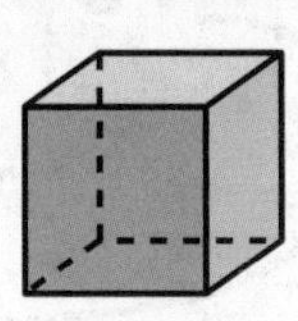

cube   rectangular prism

_______ faces   _______ vertices

4. 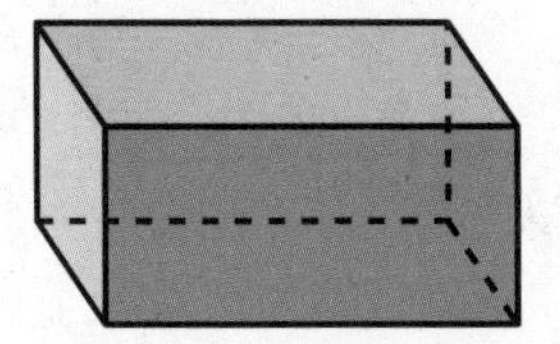

cube   rectangular prism

_______ faces   _______ vertices

Circle the shape of the face that make each object.

5. 

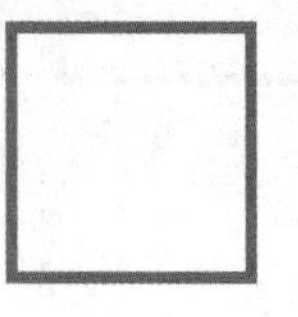 

6. 

 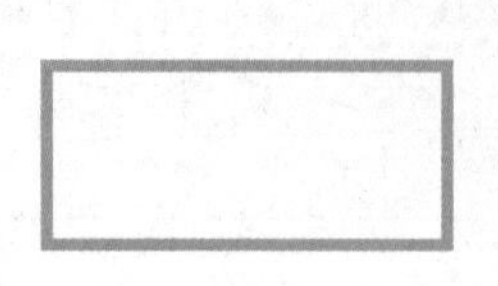

Circle the object that can be made by the faces.

7. 

Problem Solving

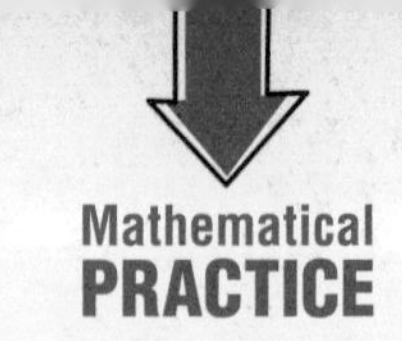

8. What shape are the faces on this cube of cheese? Draw each face.

____

9. If you put these shapes together, what three-dimensional shape do you make? Circle the name of the shape.

cube rectangular prism

Write Math How are cubes and rectangular prisms different?

Name ....

My Homework

Lesson 1
Cubes and Prisms

Homework Helper

Need help? connectED.mcgraw-hill.com

A cube and a rectangular prism are two types of three-dimensional shapes with faces and vertices.

cube

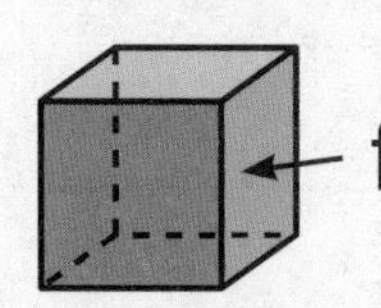

6 faces
8 vertices

rectangular prism

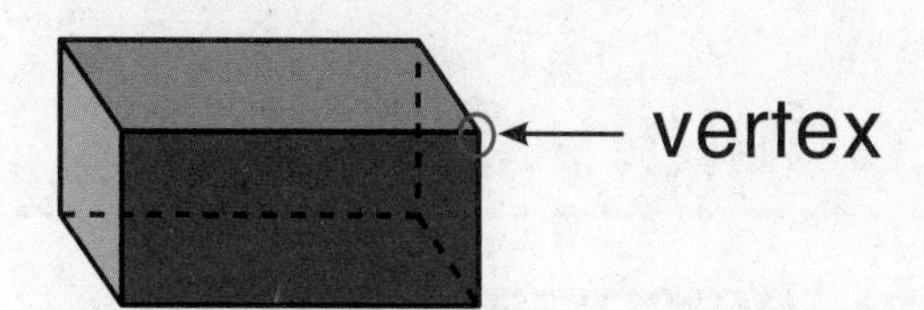

6 faces
8 vertices

Practice

Identify each shape. Circle the name. Write the number of faces and vertices.

1. 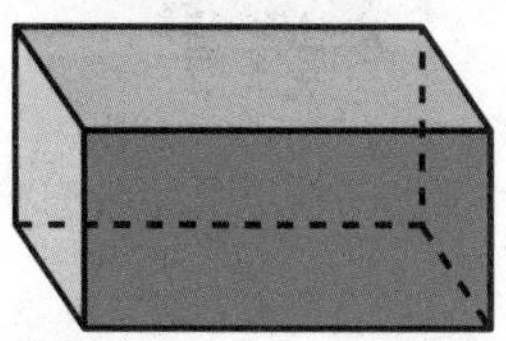

cube rectangular prism

_______ faces _______ vertices

2. 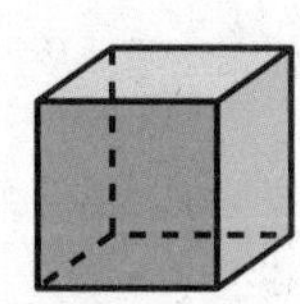

cube rectangular prism

_______ faces _______ vertices

Circle the shape of the faces that make the object.

3. 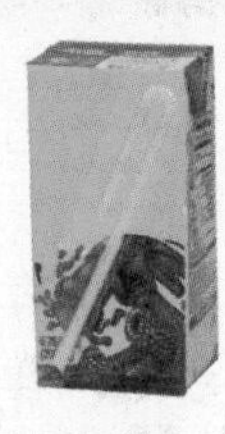

4. Kenley is wrapping a present. The present has 6 rectangular faces. What shape is the box?

____

Vocabulary Check 

Circle the correct answer.

5. rectangular prism

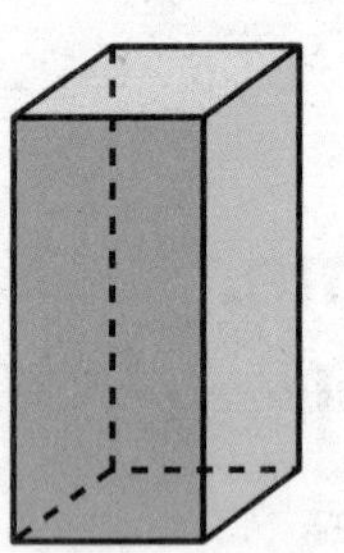

6. cube

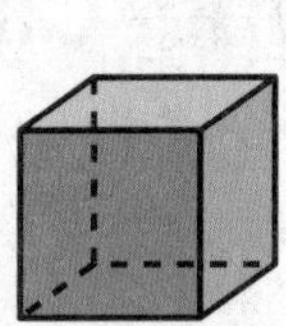 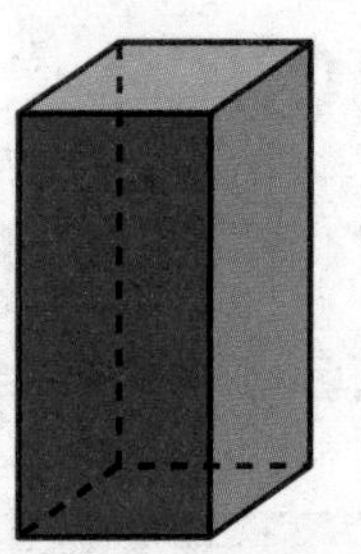

Math at Home Have your child identify and describe cubes and rectangular prisms in your home.

Name ....

Cones and Cylinders

Lesson 2

ESSENTIAL QUESTION
How can I identify three-dimensional shapes?

What's your favorite flavor?

Explore and Explain

Watch

Tools

Teacher Directions: Use and . Compare and describe the shapes. Trace one face of each shape. Explain what you notice about the faces of the cone and cylinder.

See and Show

Cones and cylinders are two more types of three-dimensional shapes. Both shapes have at least one face. Only cones have a vertex.

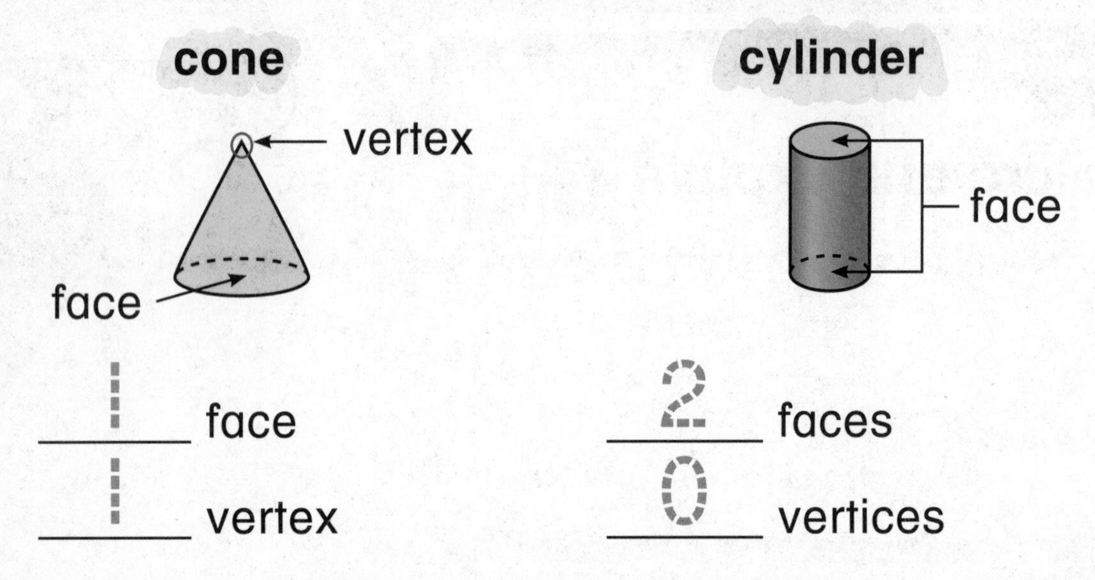

Identify each shape. Circle the name.
Write the number of faces and vertices.

1. 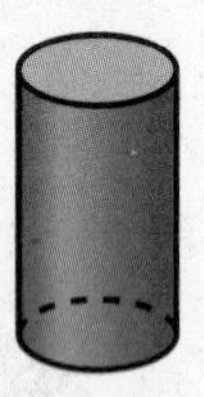

cone cylinder

_______ faces _______ vertices

2. 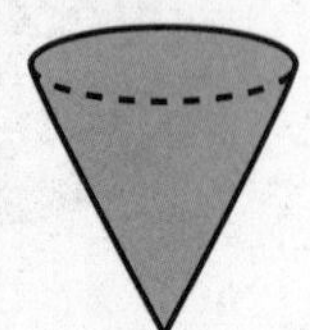

cone cylinder

_______ face _______ vertex

Talk Math How are a cone and a cylinder different?

Name ....

On My Own

Identify each shape. Circle the name.
Write the number of faces and vertices.

3. 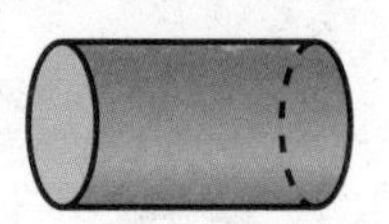

cone cylinder

_______ faces _______ vertices

4. 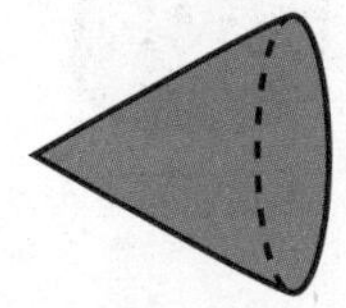

cone cylinder

_______ face _______ vertex

Circle the shape of the faces that are part of each object.

5. 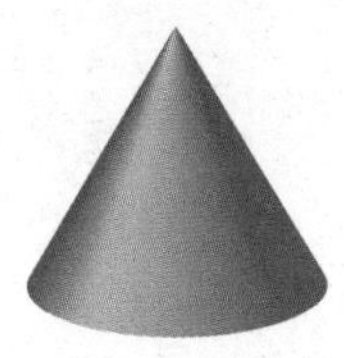

  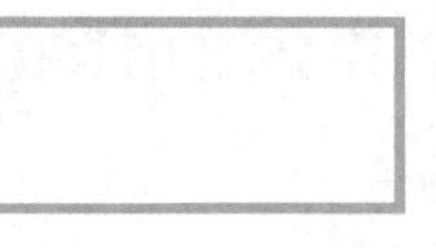 

6.

Circle the object that has the faces shown.

7. 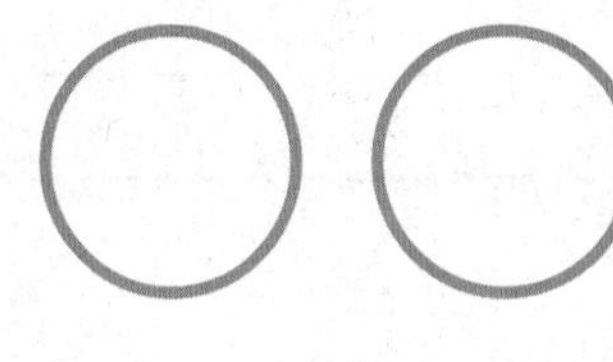

Problem Solving

8. Which three-dimensional shape has only one face?

____

9. What shape are the faces on this cylinder? Draw each of the faces.

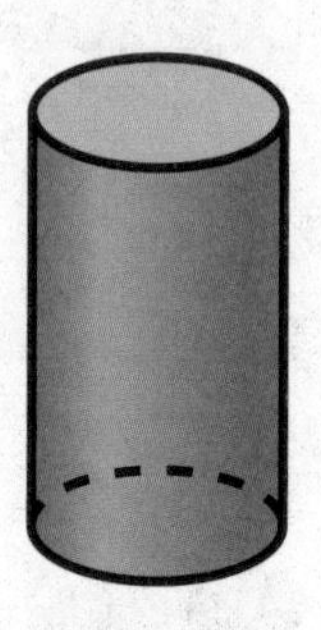

____

HOT Problem Which shape is different? Explain why it is different.

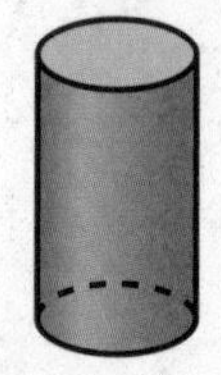 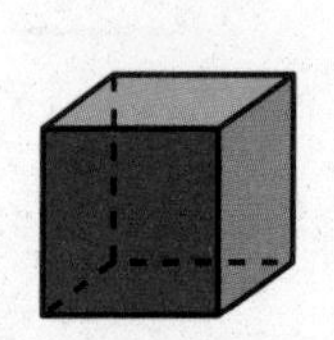 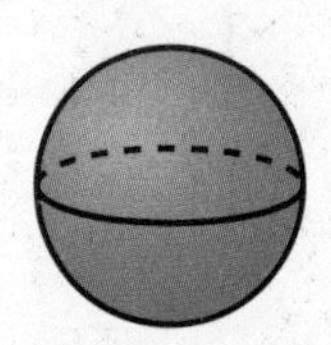 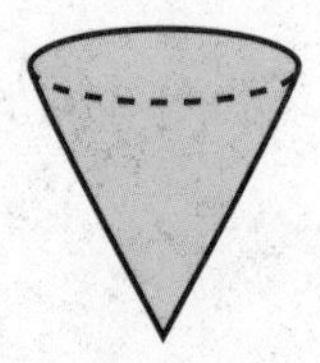

Name ....

My Homework

Lesson 2
Cones and Cylinders

Homework Helper

Need help? connectED.mcgraw-hill.com

Cones and cylinders are three-dimensional shapes. Both shapes have at least one face. Only cones have a vertex.

cone

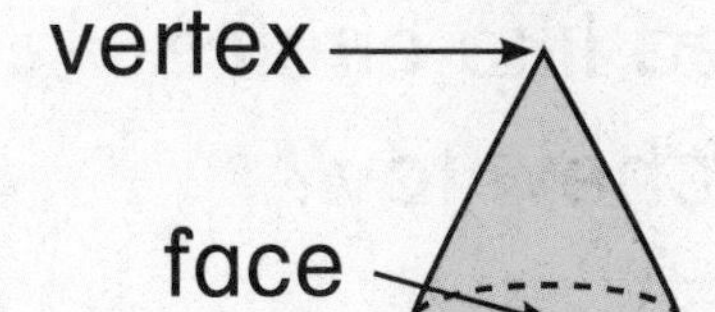

I face
I vertex

cylinder

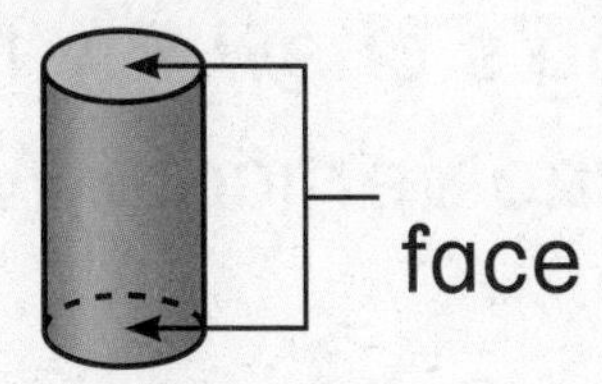

2 faces
0 vertices

Practice

Identify each shape. Circle the name.
Write the number of faces and vertices.

1. 

cone   cylinder

_______ face   _______ vertex

2. 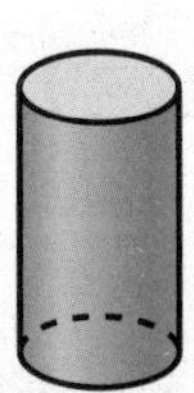

cone   cylinder

_______ faces   _______ vertices

Circle the shape of the faces that are a part of the object.

3.

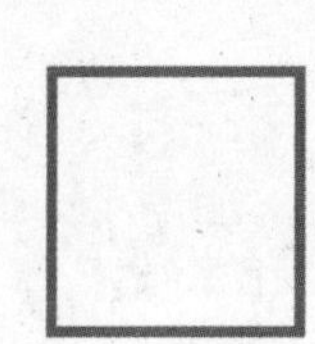

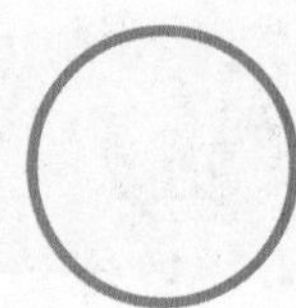

Circle the object that has the face shown.

4.

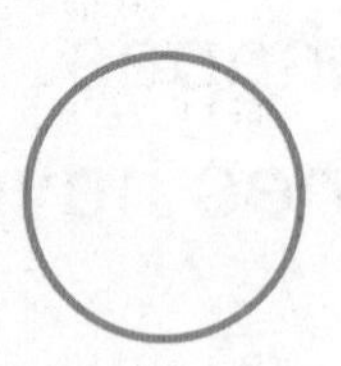

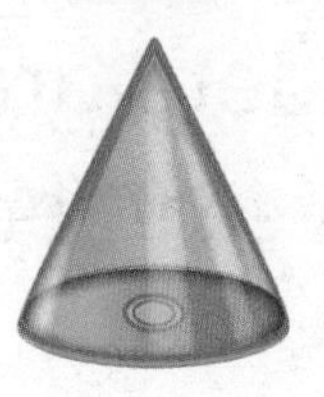

5. A dog's chew toy has 2 faces shaped like circles and no vertices. What shape is the chew toy?

____

Vocabulary Check

Circle the correct answer.

6. **cylinder**

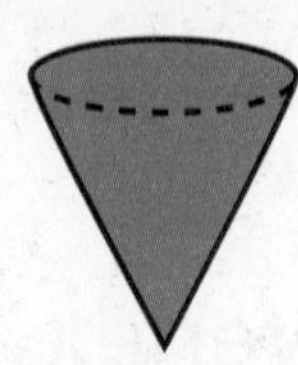

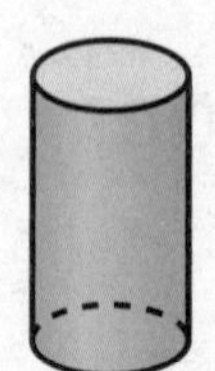

7. **cone**

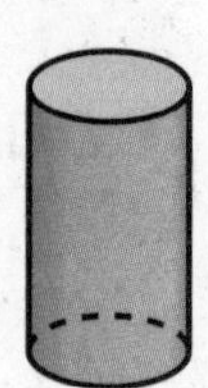

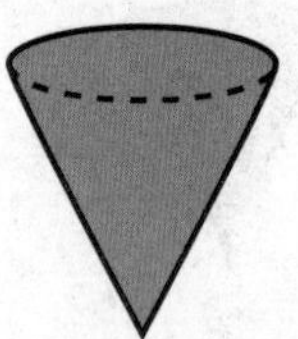

Math at Home Have your child identify and describe cones and cylinders in your home.

Name ______________________________

Check My Progress

Vocabulary Check 

Complete each sentence.

cone **cube** **cylinder** **face**

1. A ____________________ has 1 face and 1 vertex.
2. A ____________________ has 6 square faces and 8 vertices.
3. A ____________________ is the flat part of a three-dimensional shape.
4. A ____________________ has 2 faces and 0 vertices.

Concept Check

Identify each shape. Circle the name.
Write the number of faces and vertices.

5. 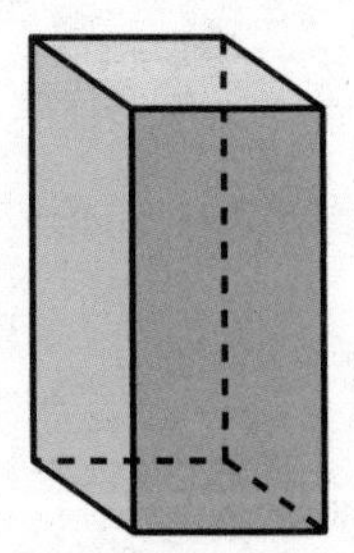

cone rectangular prism

_______ faces _______ vertices

Circle the shape of the faces that make each object.

6. 

7. 

Circle the object that can be made by the faces.

8. 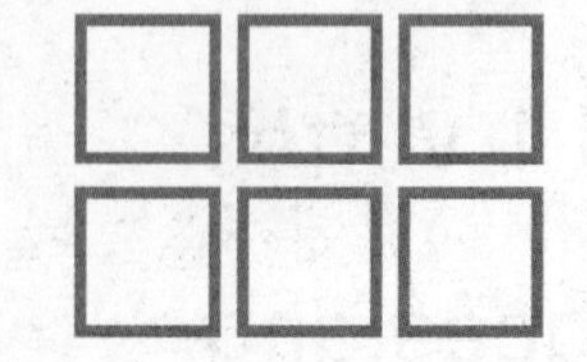

9. 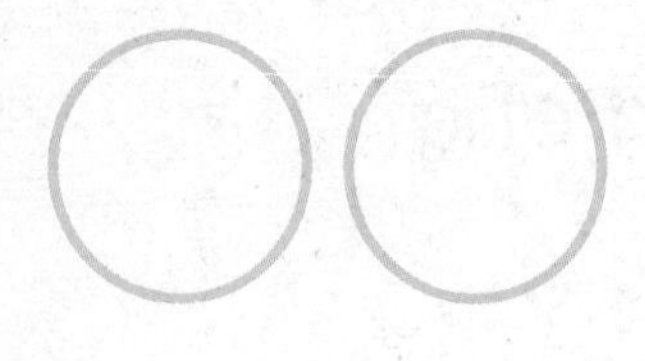

Test Practice

10. Lance took apart a cube. How many squares did he have?

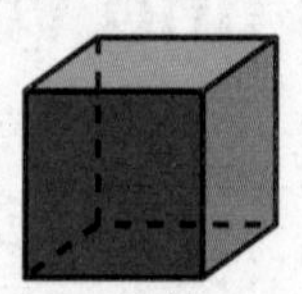

2	4	6	8
○	○	○	○

Name

Problem Solving

STRATEGY: Look for a Pattern

Lesson 3

ESSENTIAL QUESTION
How can I identify three-dimensional shapes?

Otis made a pattern with these shapes. Which shape is missing?

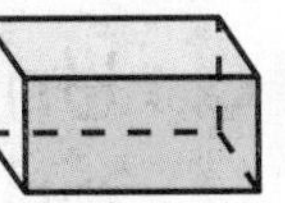

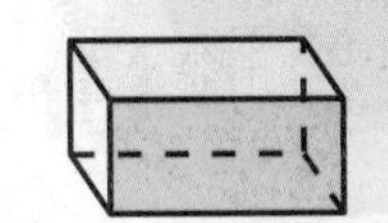

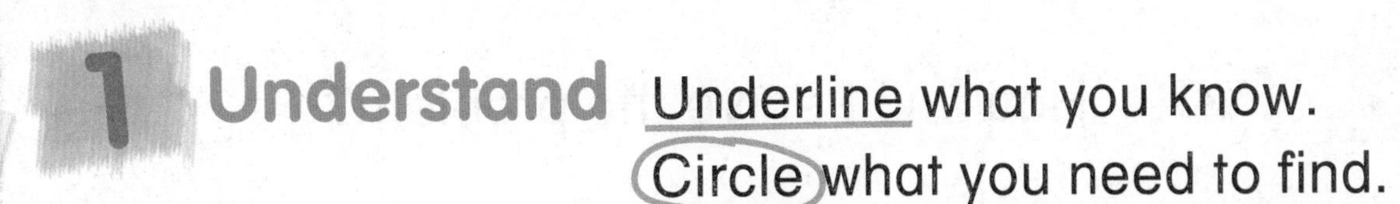

1 Understand Underline what you know.
Circle what you need to find.

2 Plan How will I solve the problem?

3 Solve I will find a pattern.

Circle the shape that is missing.

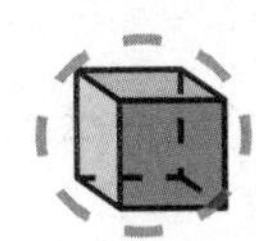
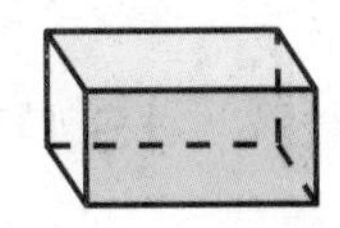

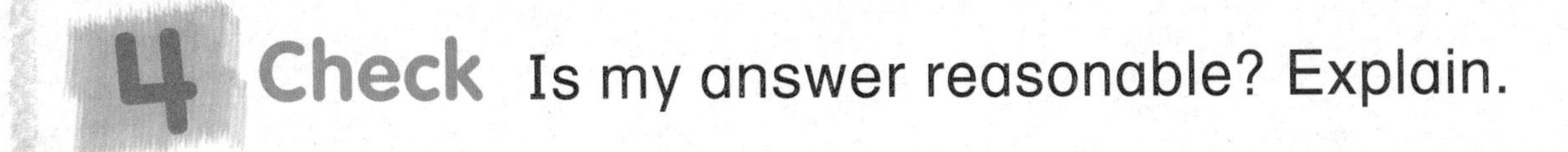

4 Check Is my answer reasonable? Explain.

Practice the Strategy

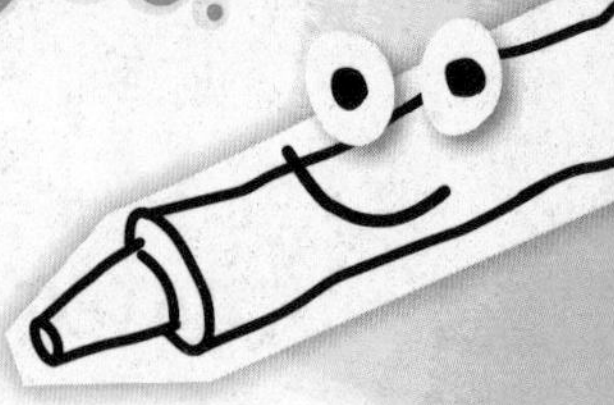

Sophie made a pattern with these shapes. Which shape comes next?

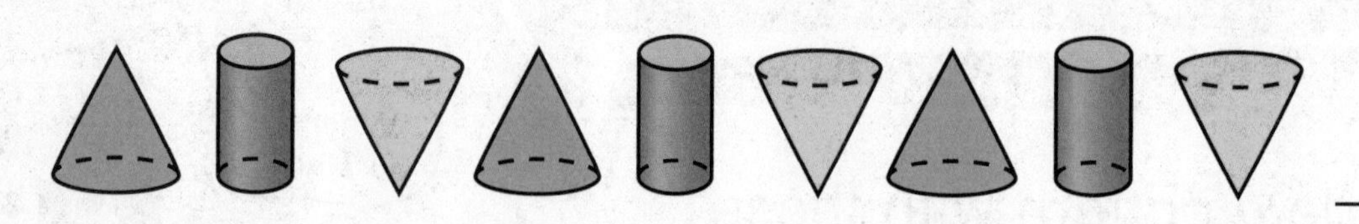

1 Understand Underline what you know.
Circle what you need to find.

2 Plan How will I solve the problem?

3 Solve I will . . .

Circle the shape that comes next.

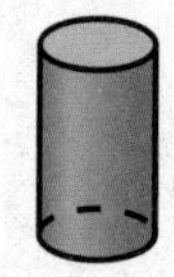

4 Check Is my answer reasonable? Explain.

Name

Apply the Strategy

Find a pattern to solve.

1. June made this pattern.

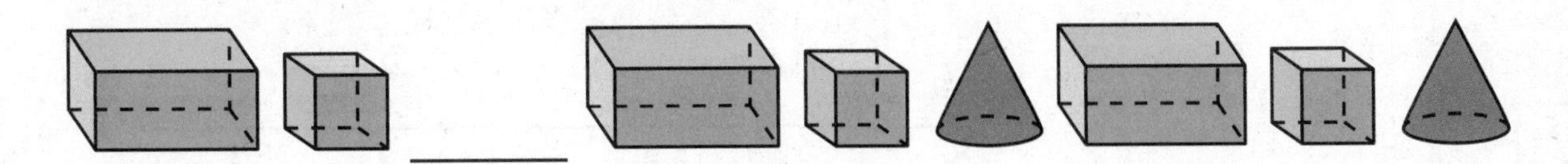

What shape is missing? Circle it.

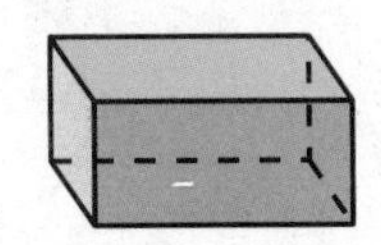  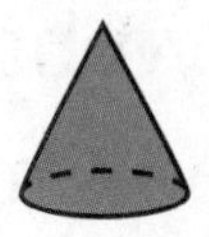

2. These are Quinn's blocks.

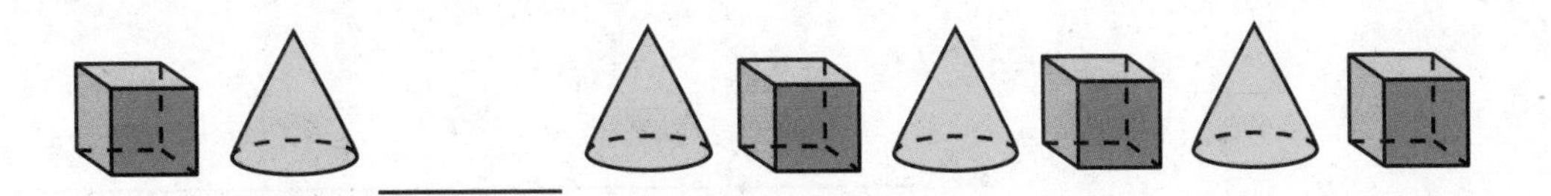

What shape is missing? Circle it.

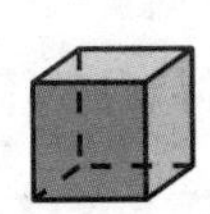 

3. Ayden made a line of blocks.

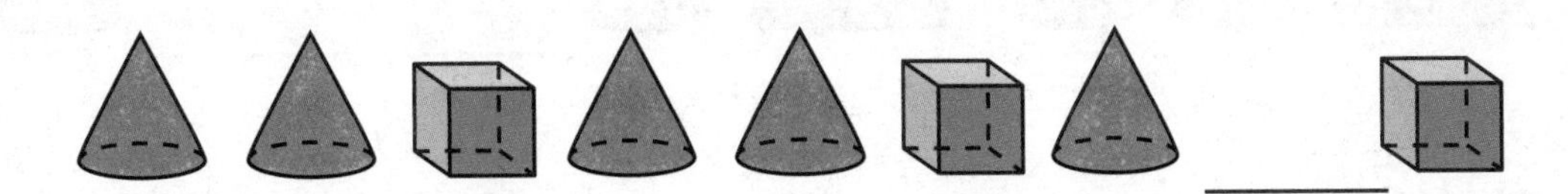

What shape is missing? Circle it.

cone cube

Review the Strategies

Choose a strategy

- Find a pattern.
- Draw a diagram.
- Use logical reasoning.

4. I have one face. I have one vertex. What shape am I?

____

5. LaToya buys a box of tissues. The box has 8 vertices. All of the faces are rectangles. What shape is the box of tissues?

____

6. Grace made the row of shapes shown. She needs 2 more blocks to finish the pattern.

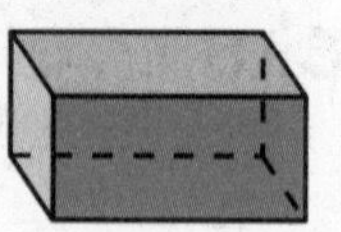

____ ____

What two shapes does she need? Circle them.

cube　　cone　　rectangular prism

Name ....

My Homework

Lesson 3
Problem Solving: Look for a Pattern

Homework Helper

 Need help? connectED.mcgraw-hill.com

Arnold made a pattern with these blocks.
What shape comes next?

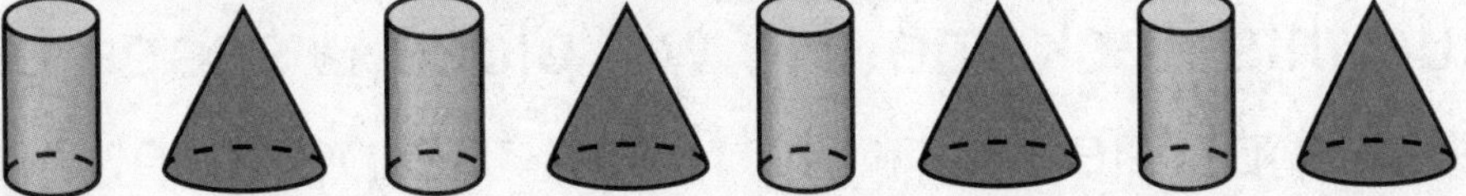
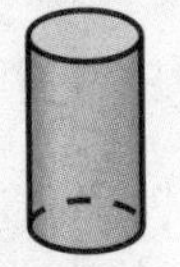

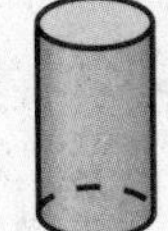
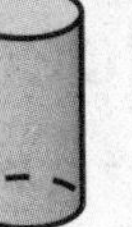
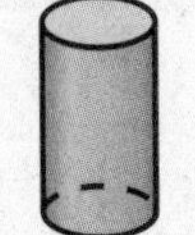
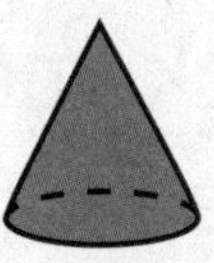

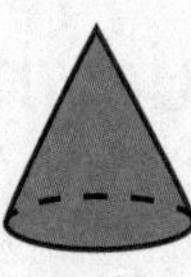

______

1 Understand Underline what you know.
Circle what you need to find.

2 Plan How will I solve the problem?

3 Solve I will find a pattern.

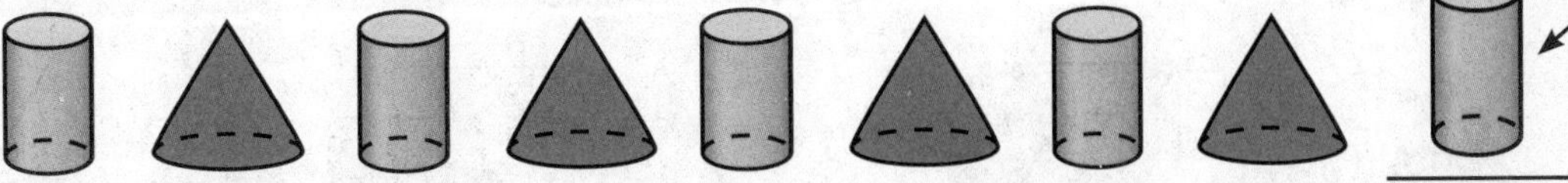

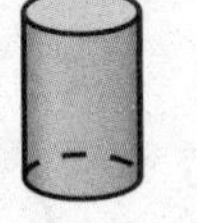

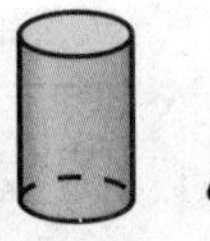
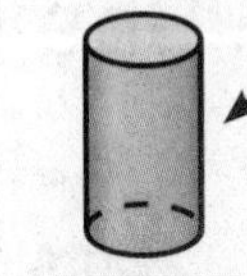

The next shape is a cylinder.

4 Check Is my answer reasonable?

Problem Solving

Find a pattern to solve.

1. Jenica made this pattern.

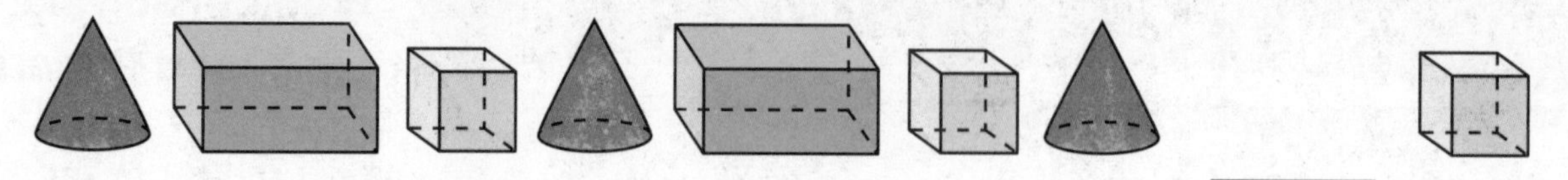

What shape is missing? Circle it.

 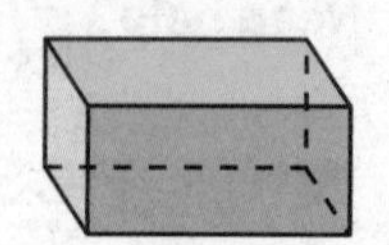 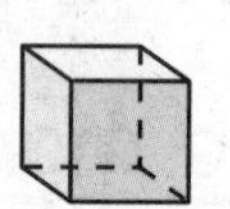

2. Alex made this necklace. He will place a shape on the right side of the string to finish the pattern.

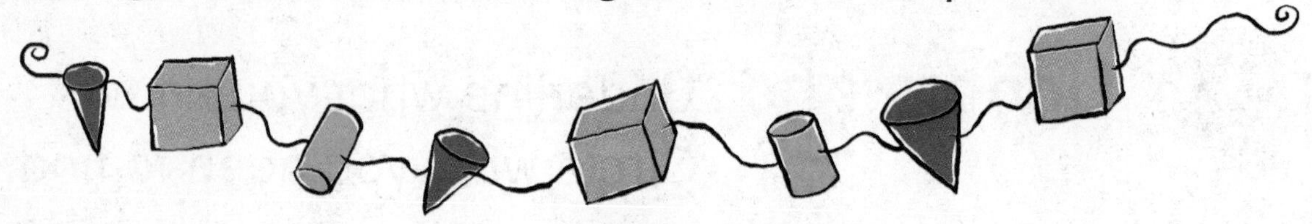

What shape is missing? Circle it.

3. Julie made a row of shapes. She left out one shape.

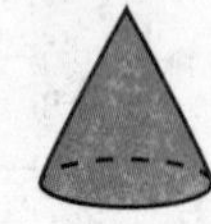  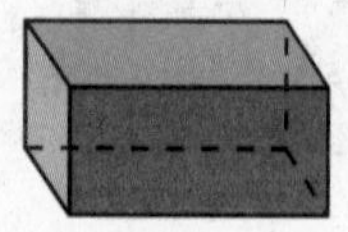 ______ 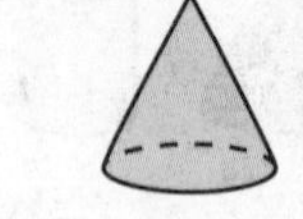 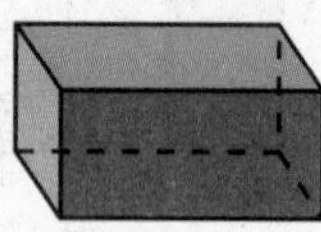 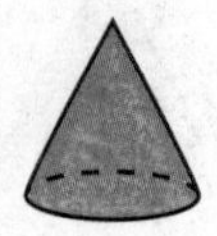  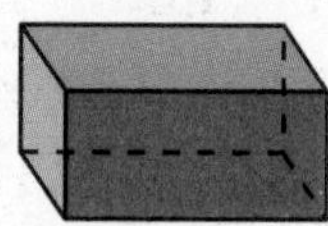

What shape is missing? Circle it.

blue cone   yellow cone   red rectangular prism

Math at Home Create a pattern out of objects that are three-dimensional shapes. Have your child copy the pattern.

Name

Combine Three-Dimensional Shapes

Lesson 4

ESSENTIAL QUESTION
How can I identify three-dimensional shapes?

Explore and Explain

Can I join the fun?

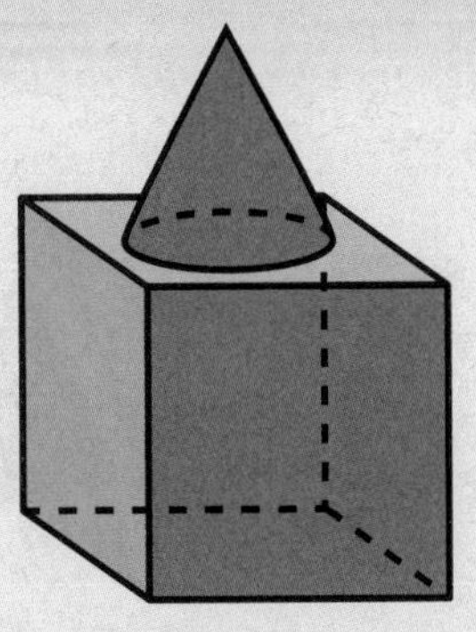

cube rectangular prism cone

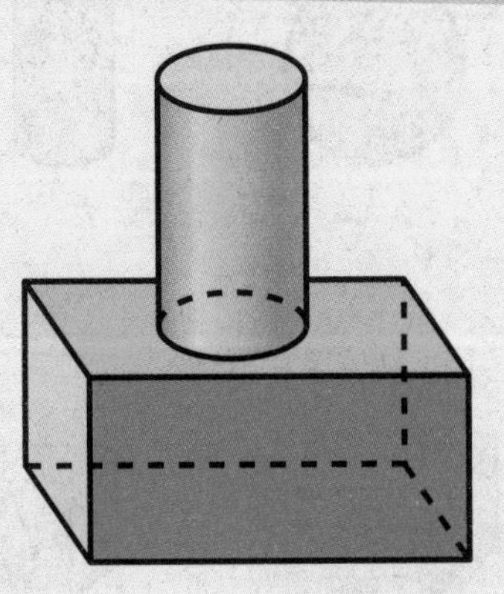

rectangular prism cylinder cube

Teacher Directions: Use geometric solids to make each composite shape shown. Circle the name of the shapes you used to make the composite shape.

See and Show

You can put together three-dimensional shapes to make other composite shapes.

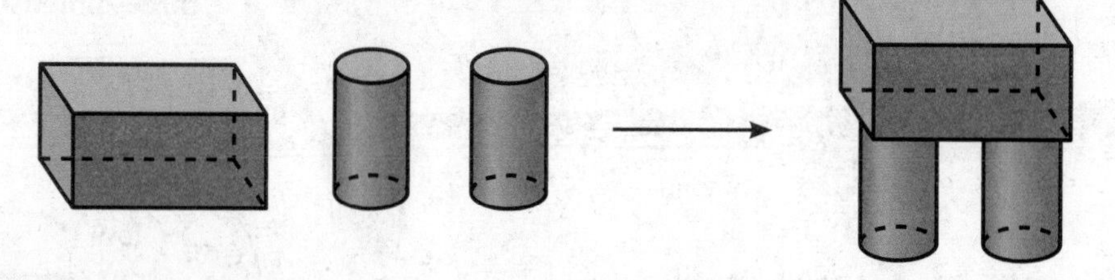

Circle the shapes used to make each composite shape.

1. 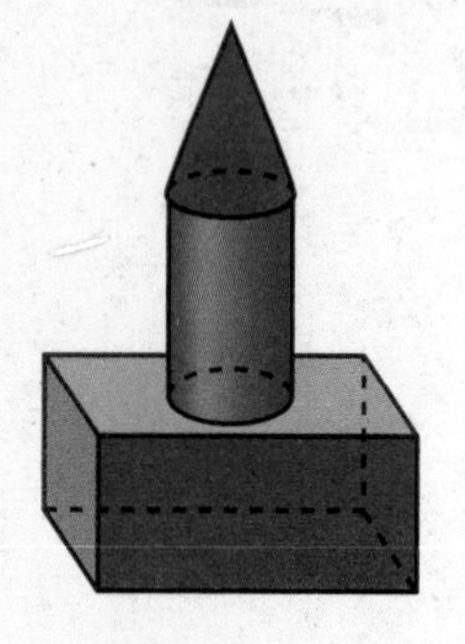 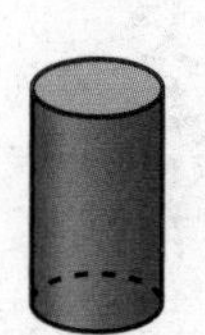 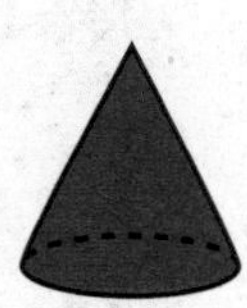 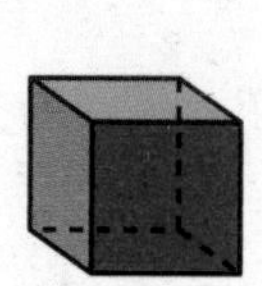 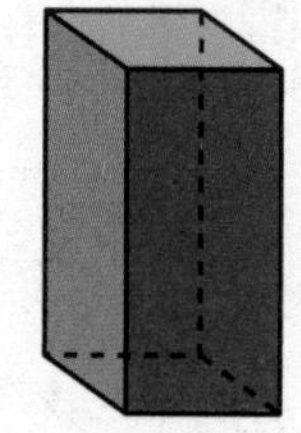

2. 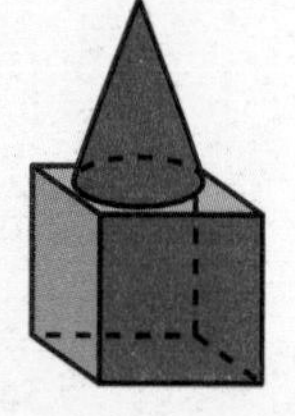 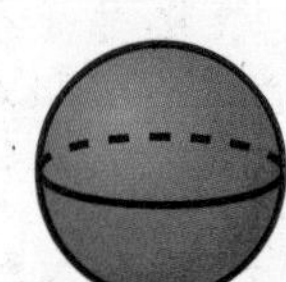 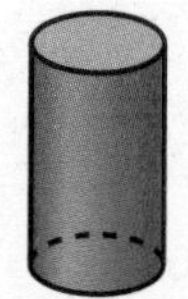 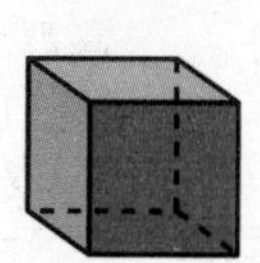 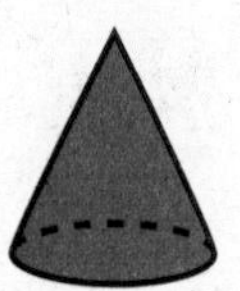

3. 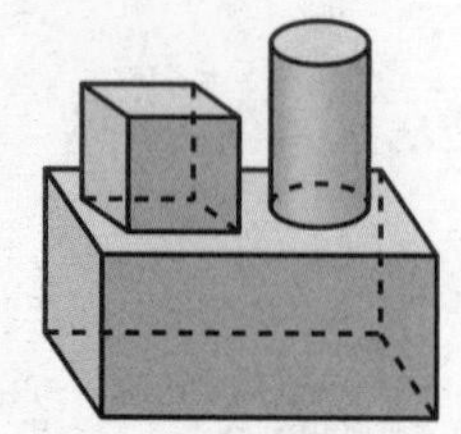 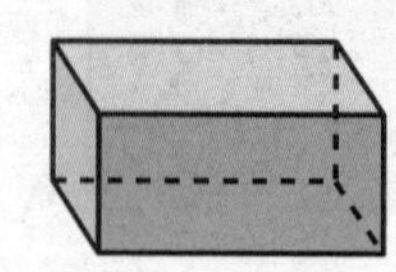 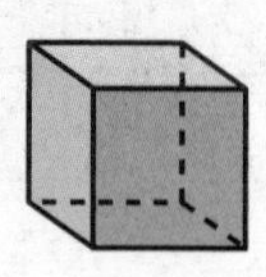 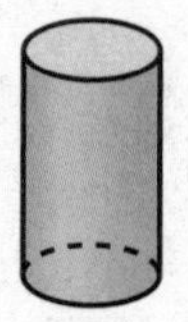 

Talk Math Will a cube stack on top of a sphere?

Name

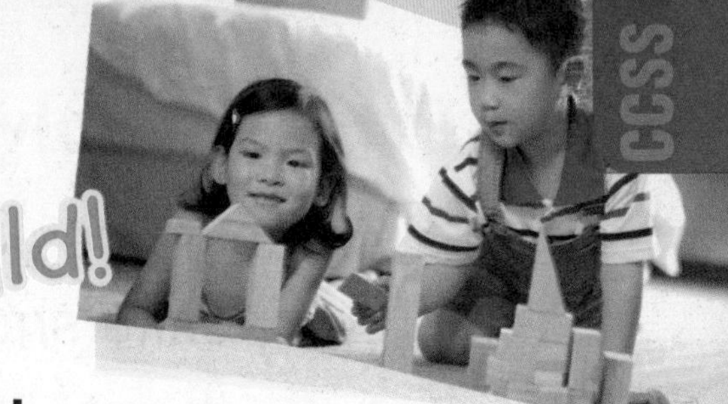

On My Own

Let's build!

Circle the shapes used to make each composite shape.

4. 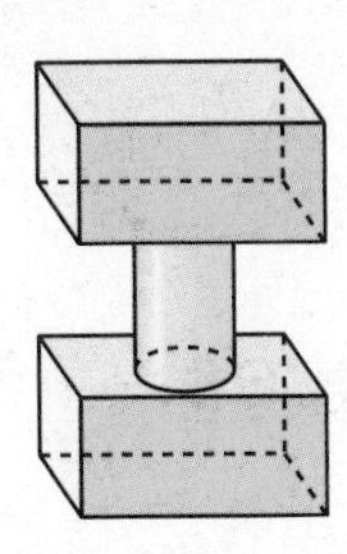 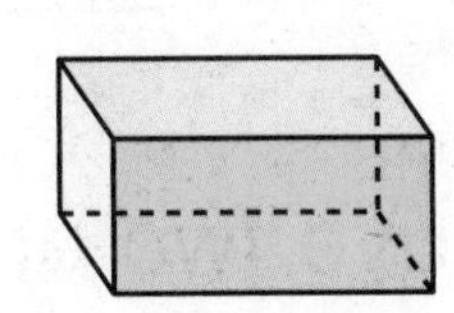 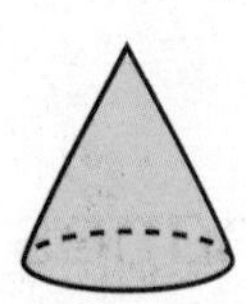 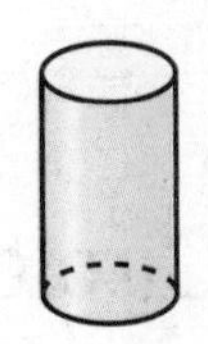

5. 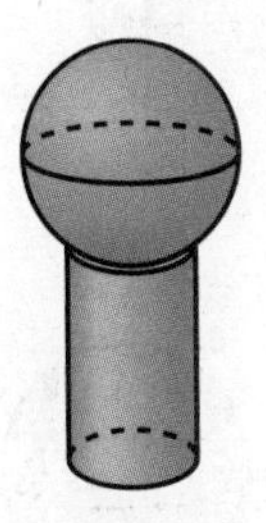 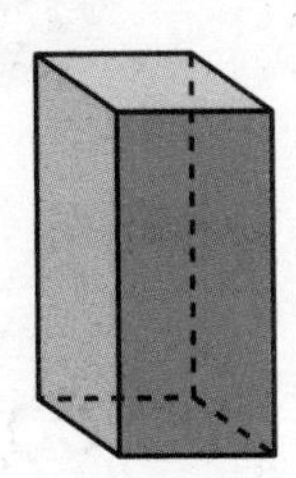 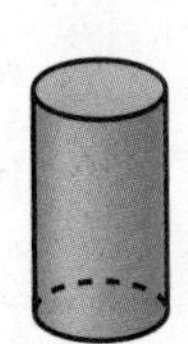 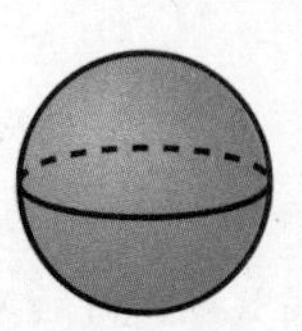

6. 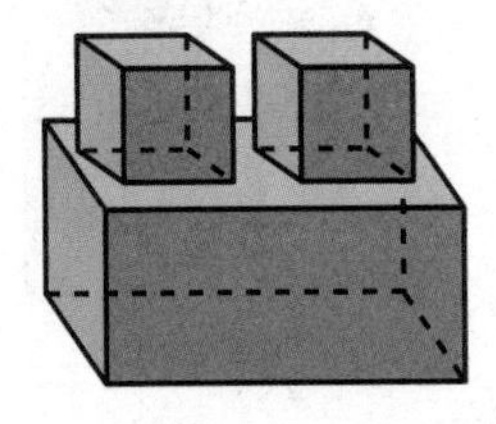 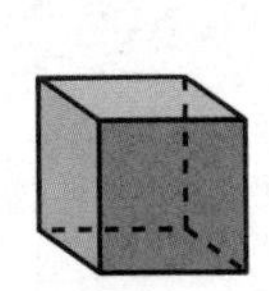 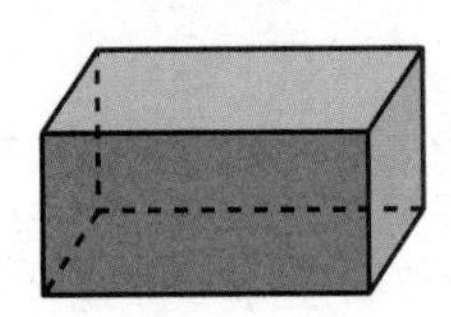 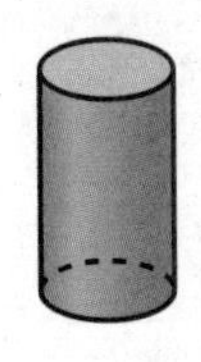

Circle the shape that is not used to make the composite shape shown.

7. 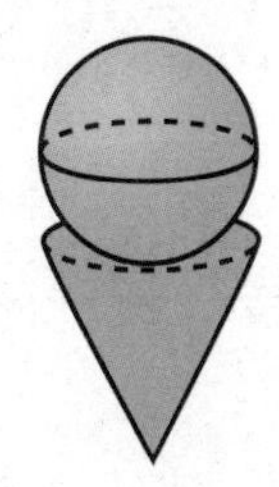 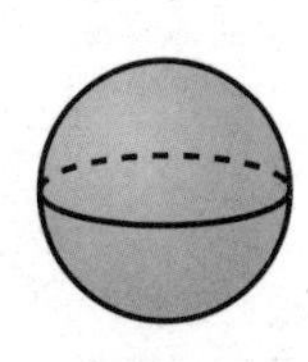 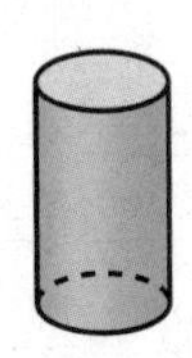 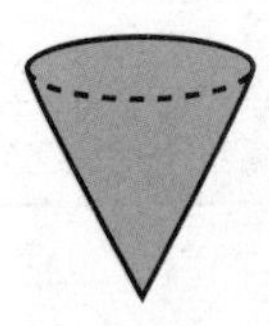

Problem Solving

8. How many faces are there in all on the composite shape shown?

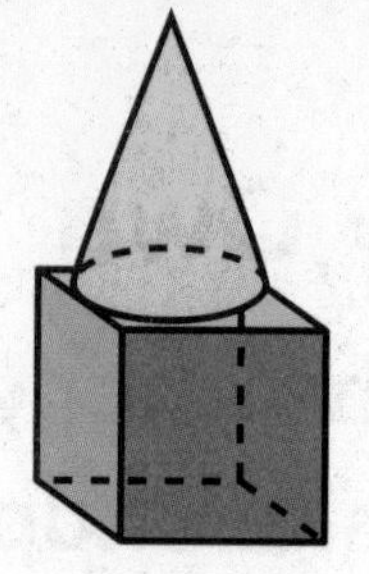

_____ faces

9. Circle the shapes that have two or more faces.

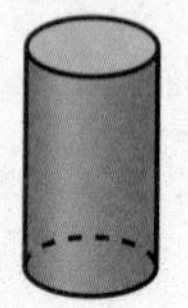 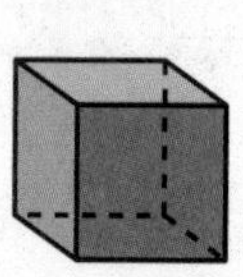 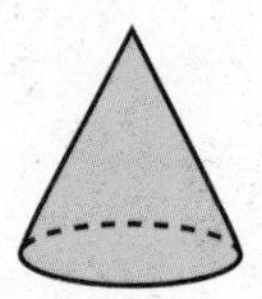 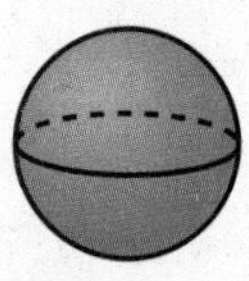 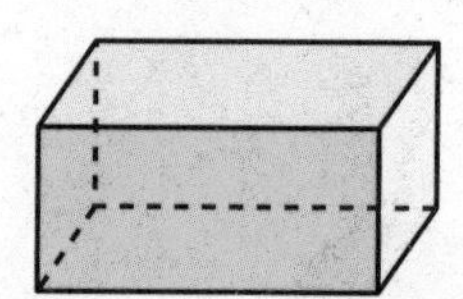

HOT Problem Angie built this composite shape. Describe another composite shape she can build using these shapes.

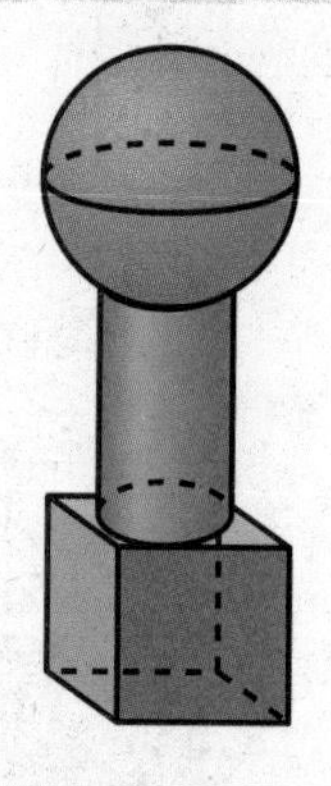

Name ....

Geometry
1.G.2
CCSS

My Homework

Lesson 4
Combine Three-Dimensional Shapes

Homework Helper

eHelp Need help? connectED.mcgraw-hill.com

You can put together three-dimensional shapes to make other composite shapes.

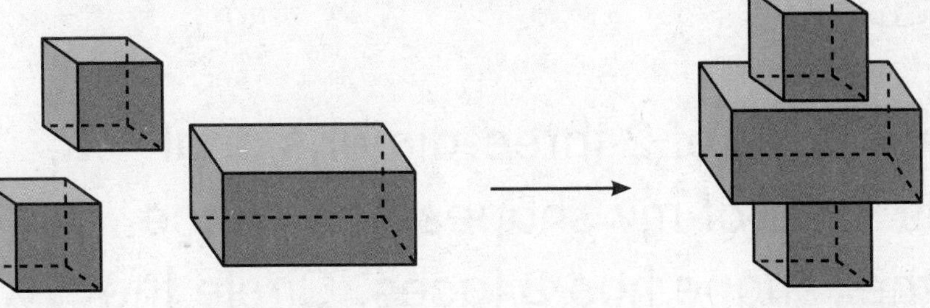

Practice

Circle the shapes used to make each composite shape.

1.

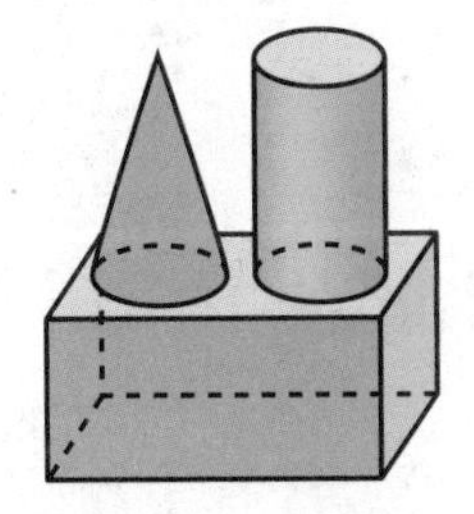

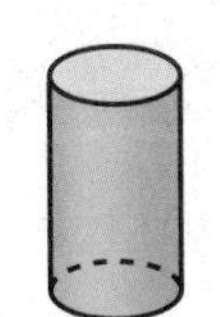

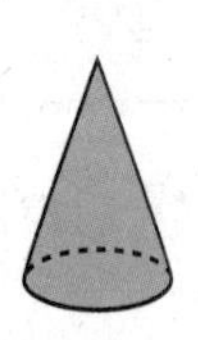

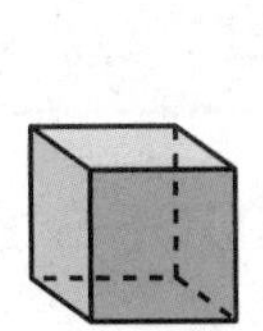

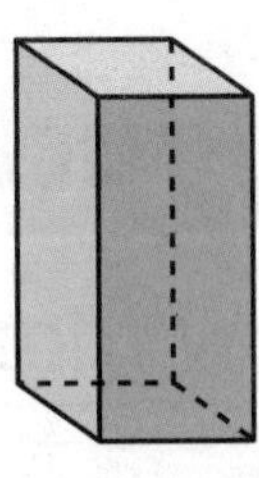

2.

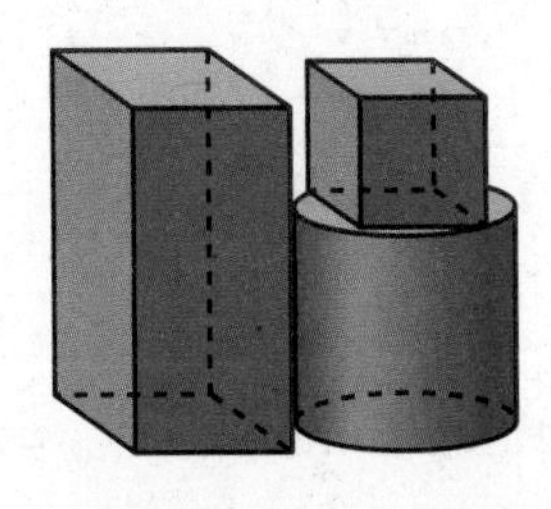

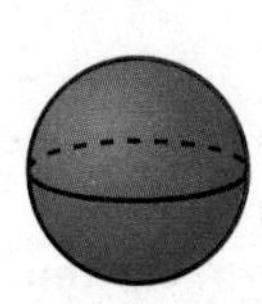

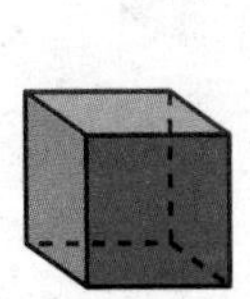

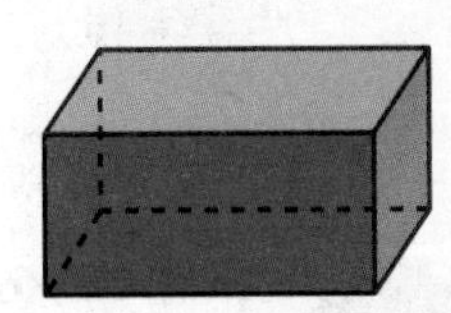

Circle the shape that is not used to make each composite shape shown.

3. 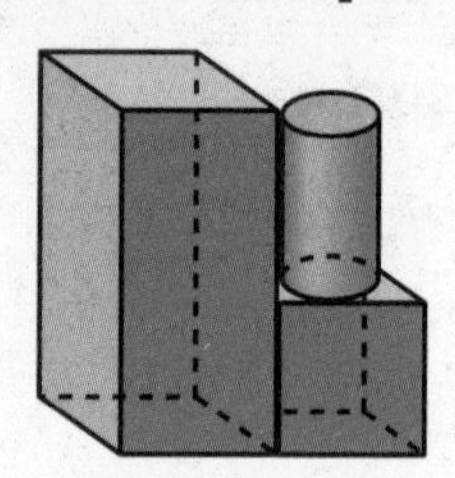

 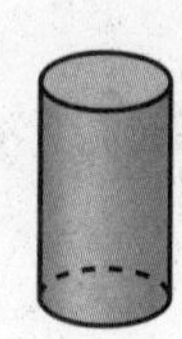 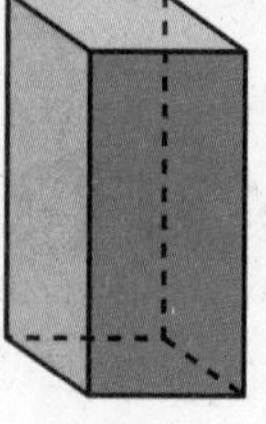 

4. 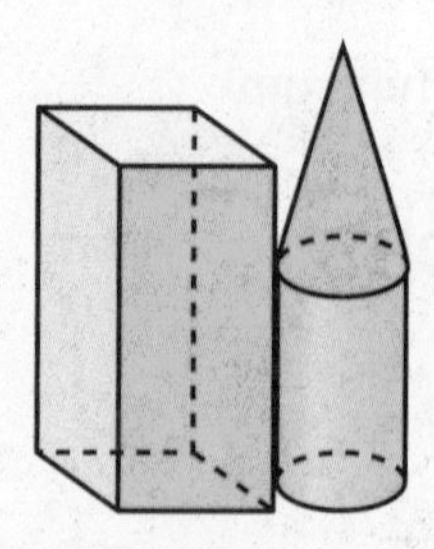

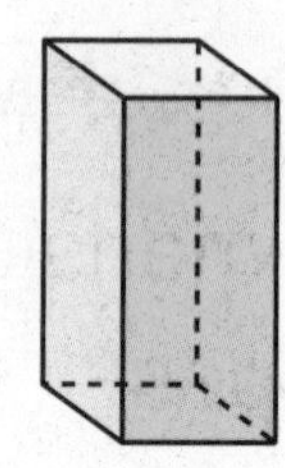 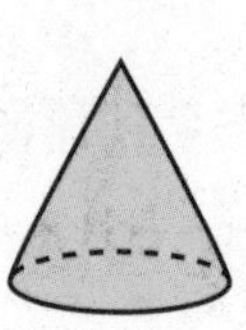 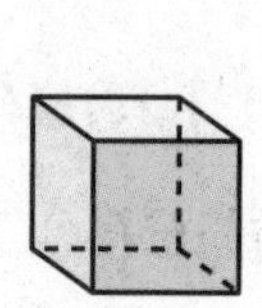 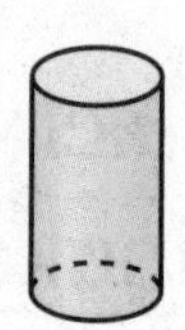

5. I am made up of 2 three-dimensional shapes. One of my shapes has 1 face. The other shape has 2 faces. Circle the 2 shapes.

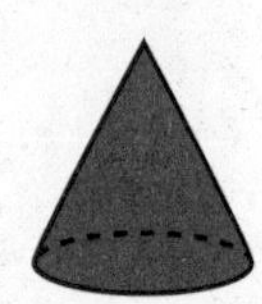 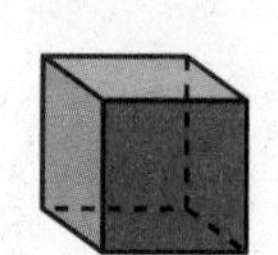 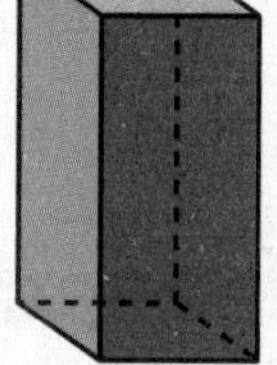 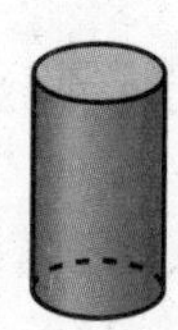

Test Practice

6. How many faces do a cube and cylinder have in all?

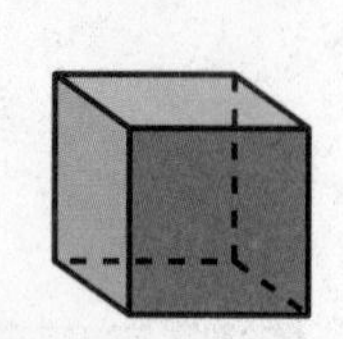 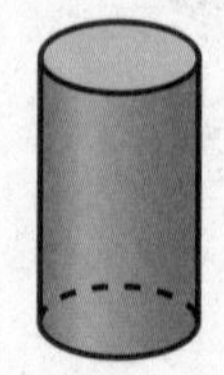

2	4	6	8
○ 	○	○ 	○ 

Math at Home Have your child find different three-dimensional objects in your house. Have them build new composite shapes with these objects. Have them tell you which shapes they used to create the new composite shapes.

Name ....

My Review

Chapter 10

Three-Dimensional Shapes

Vocabulary Check 

Draw lines to match.

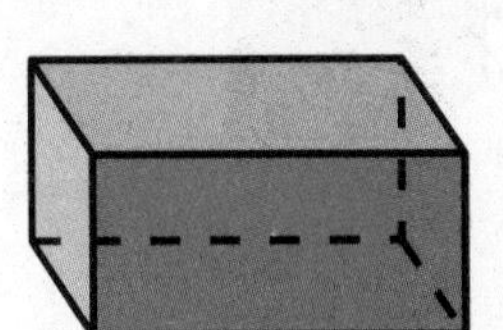

1. **cone**

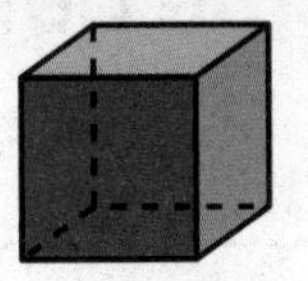

2. **cube** 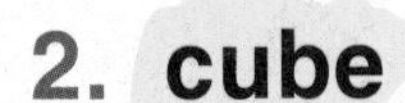

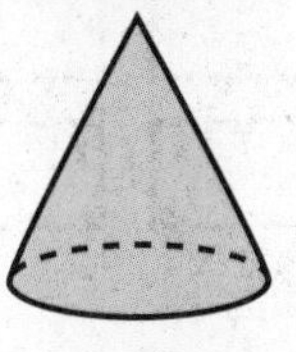

3. **cylinder** 

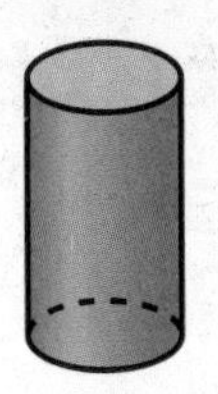

4. **rectangular prism**

Concept Check

Identify each shape. Circle the name.
Write the number of faces and vertices.

5. 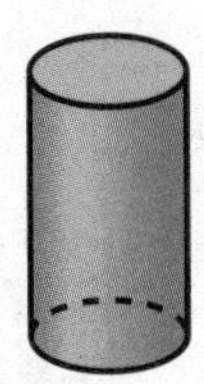

cylinder   cube

_______ faces   _______ vertices

Identify each shape. Circle the name.
Write the number of faces and vertices.

6. 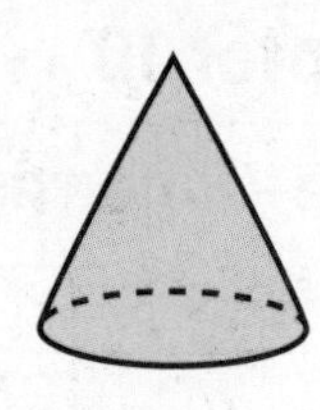

cone   rectangular prism

_______ face   _______ vertex

7. 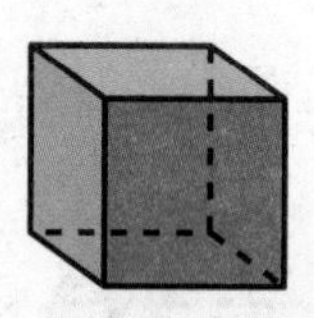

cube   cylinder

_______ faces   _______ vertices

Circle the object that can be made by the faces.

8. 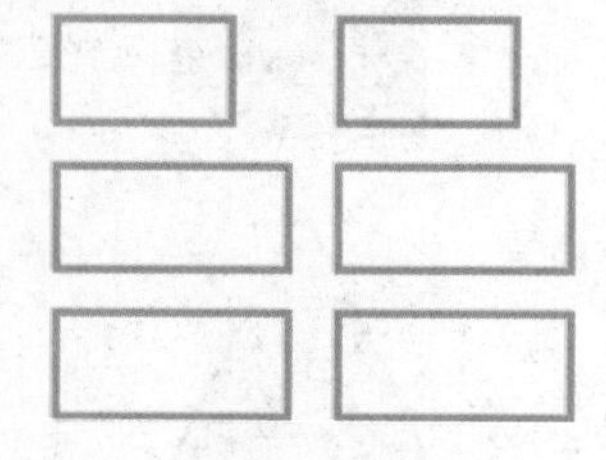

Circle the object that has the face shown.

9. 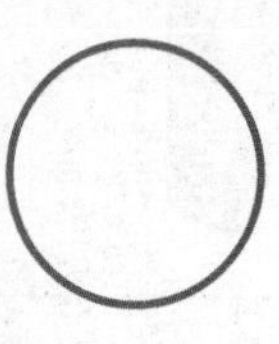

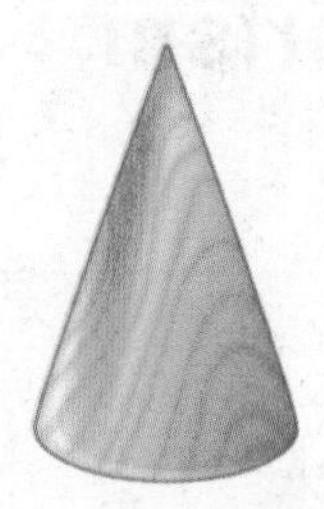

Circle the shape that is not used to build the composite shape below.

10. 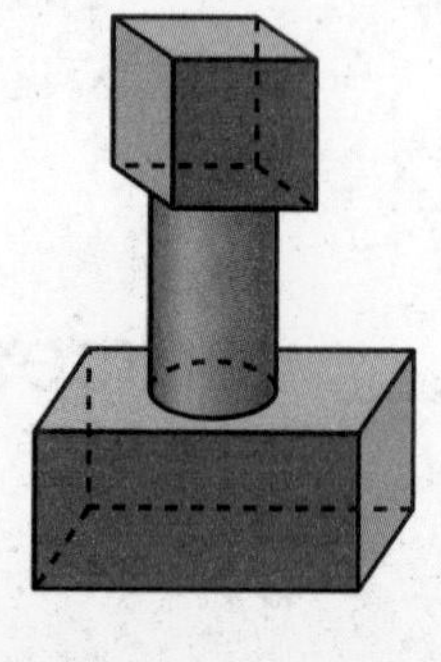

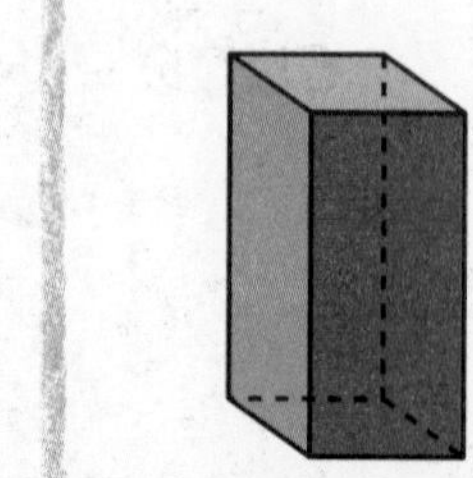

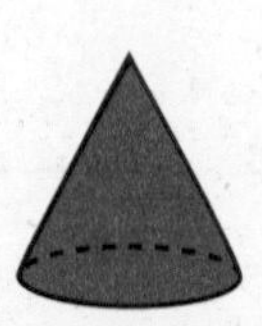

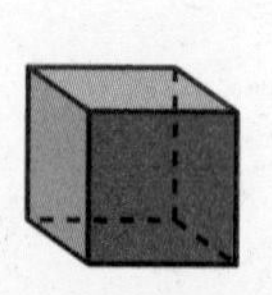

Name

Problem Solving

11. If you put these shapes together, what three-dimensional shape do you make?

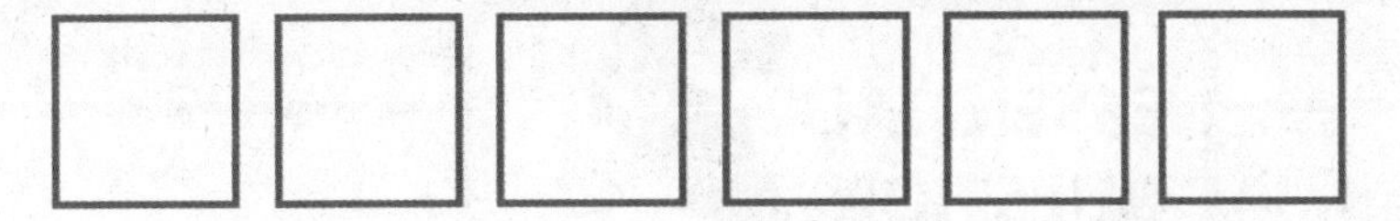

Circle the name of the shape they make.

cube rectangular prism

12. Circle the shapes with 4 or more faces.

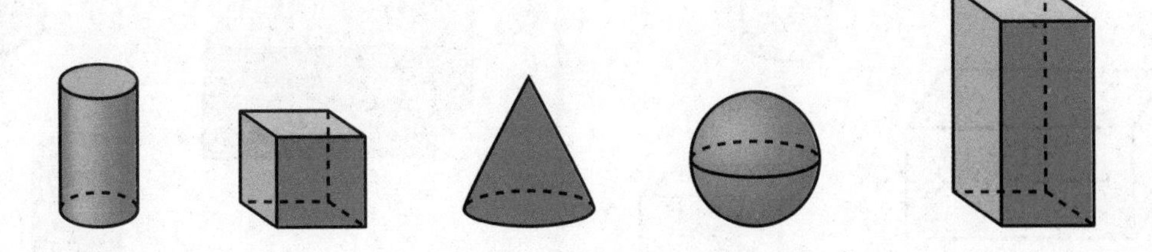

Test Practice

13. I am a three-dimensional shape. My faces are shaped like squares. I have 8 vertices. What shape am I?

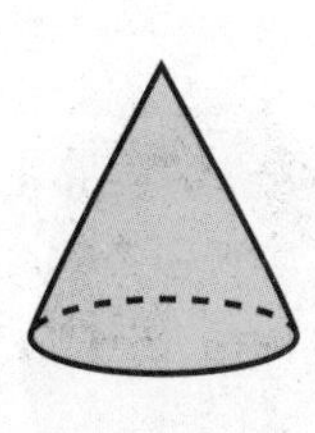
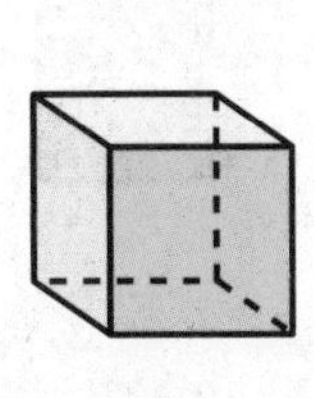
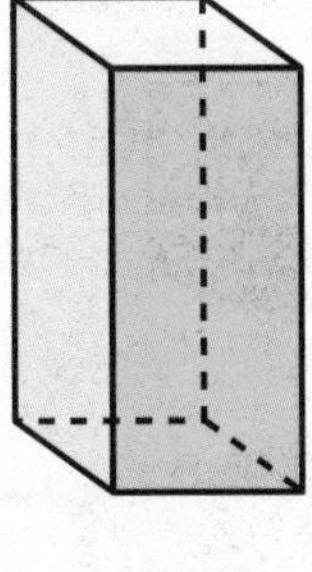
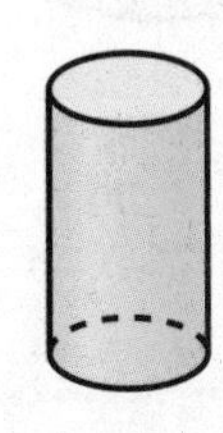

Reflect

Chapter 10
Answering the Essential Question

Circle the shapes in each box that follow the rules.

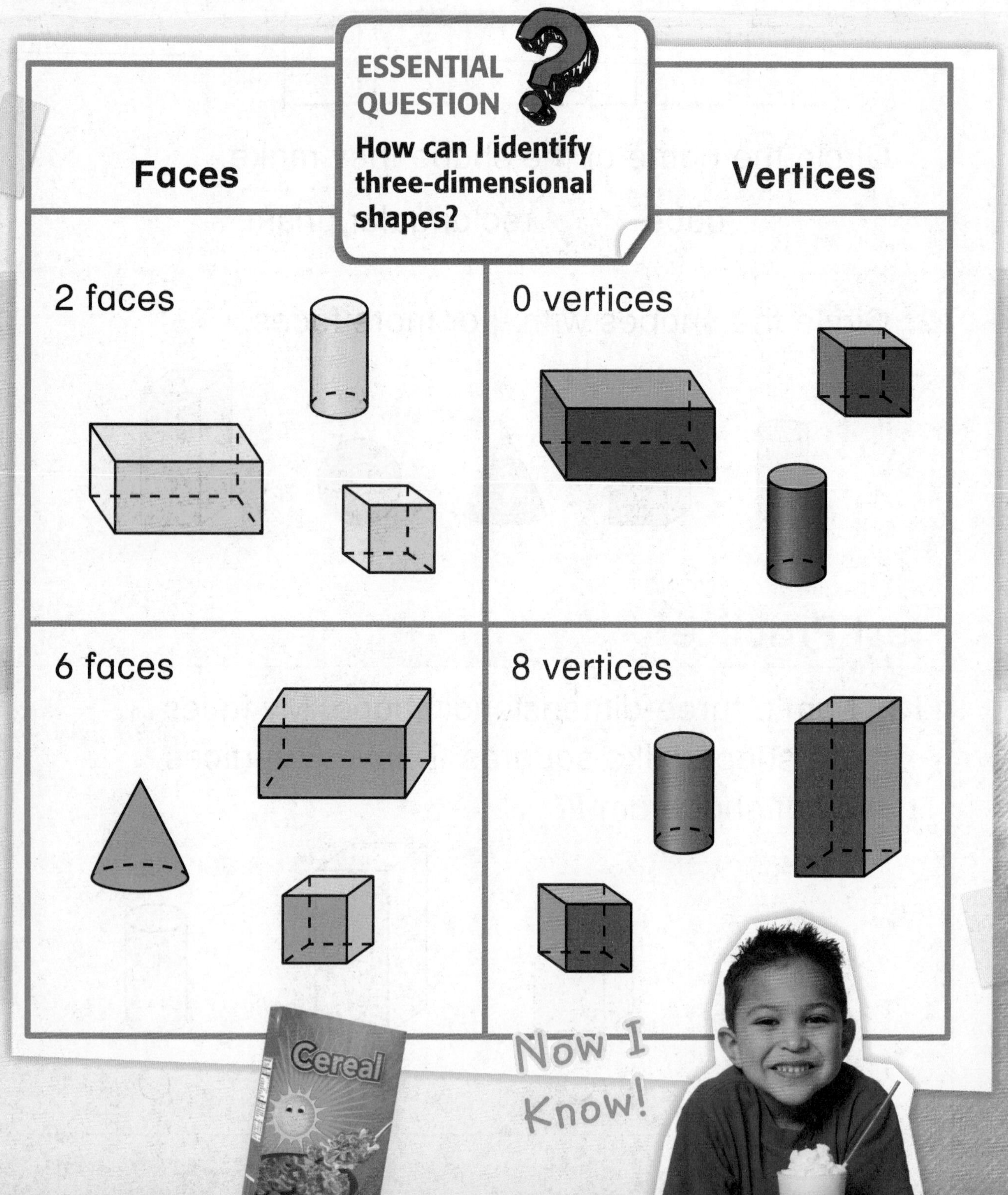

Glossary/Glosario

Go online for the eGlossary.

Aa

English | Spanish/Español

add (adding, addition) To join together sets to find the total or sum.

$2 + 5 = 7$

sumar (adición) Unir conjuntos para hallar el total o la suma.

$2 + 5 = 7$

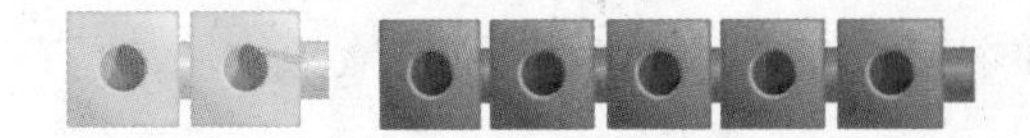

addend Any numbers or quantities being added together.

$2 + 3$

2 is an addend and 3 is an addend.

sumando Números o cantidades que se suman.

$2 + 3$

2 es un sumando y 3 es un sumando.

addition number sentence An expression using numbers and the + and = signs.

$4 + 5 = 9$

enunciado numérico de suma Expresión en la cual se usan números con los signos + e =.

$4 + 5 = 9$

Aa

after To follow in place or time.

5 6 7 8

6 is just *after* 5

después Que sigue en lugar o en tiempo.

5 6 7 8

6 está justo *después* de 5.

analog clock A clock that has an hour hand and a minute hand.

reloj analógico Reloj que tiene manecilla horaria y minutero.

Bb

bar graph A graph that uses bars to show data.

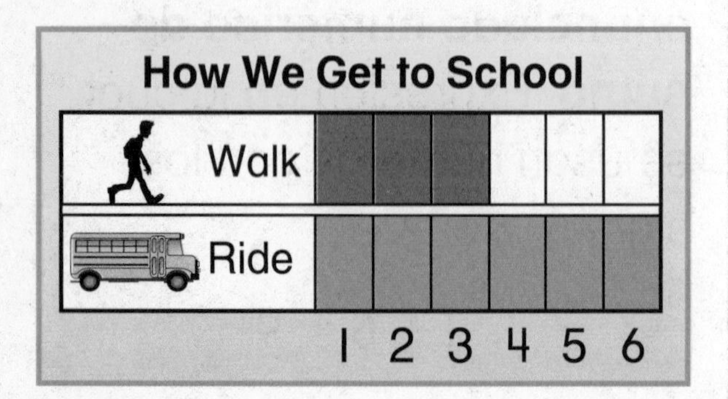

gráfica de barras Gráfica que usa barras para ilustrar datos.

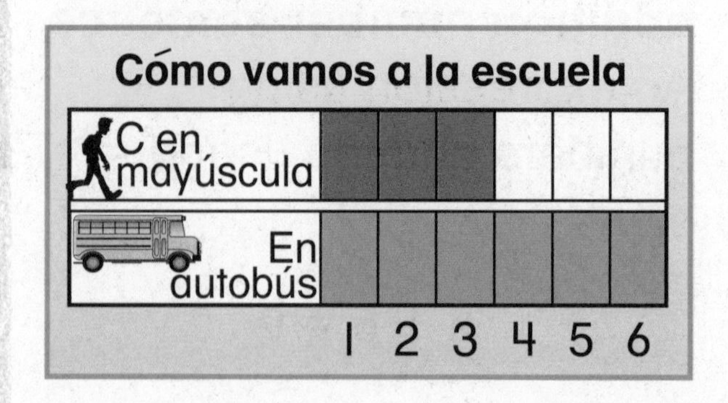

before

5 6 7 8

6 is just *before* 7.

antes

5 6 7 8

6 está justo *antes* del 7.

between

The kitten is *between* the two dogs.

entre

El gatito está *entre* dos perros.

capacity The amount of dry or liquid material a container can hold.

capacidad Cantidad de material seco o líquido que cabe en un recipiente.

Cc

cent ¢

1¢ 1 cent

centavo ¢

1¢ 1 centavo

circle A closed round shape.

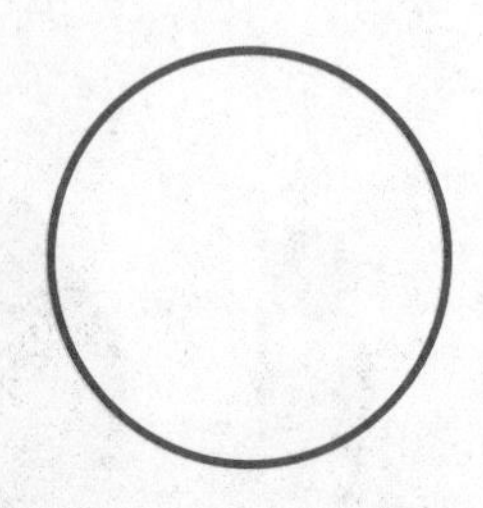

círculo Figura redonda y cerrada.

compare Look at objects, shapes, or numbers and see how they are alike or different.

comparar Observar objetos, formas o números para saber en qué se parecen y en qué se diferencian.

cone A three-dimensional shape that narrows to a point from a circular face.

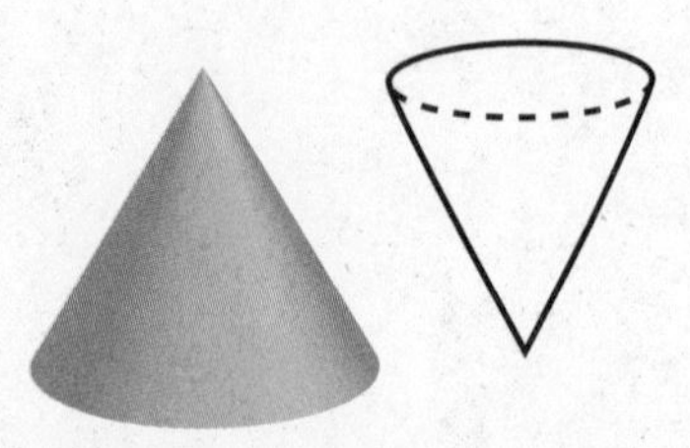

cono Una figura tridimensional que se estrecha hasta un punto desde una cara circular.

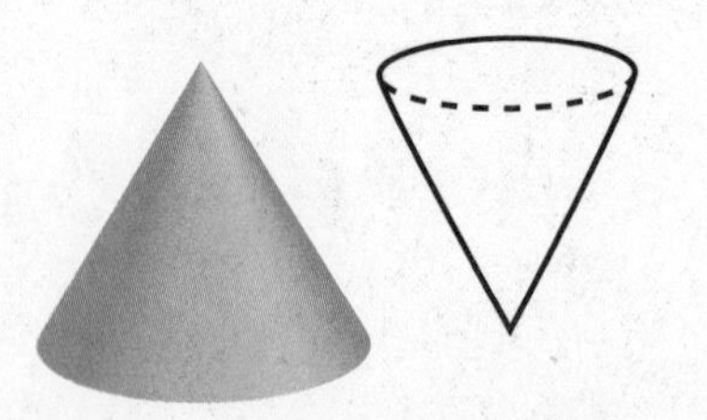

count back On a number line, start at the number 5 and count back 3.

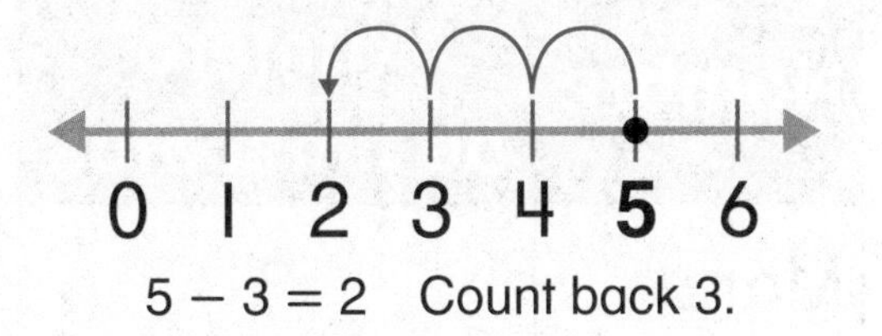

5 − 3 = 2 Count back 3.

contar hacia atrás En una recta numérica, comienza en el número 5 y cuenta 3 hacia atrás.

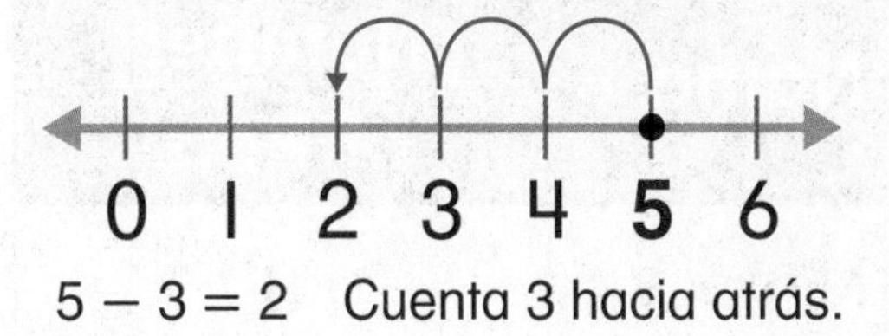

5 − 3 = 2 Cuenta 3 hacia atrás.

count on (or count up) On a number line, start at the number 4 and count up 2.

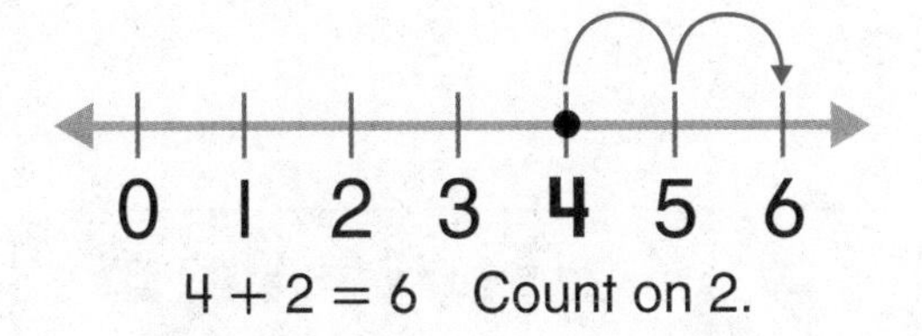

4 + 2 = 6 Count on 2.

seguir contando (o contar hacia delante) En una recta numérica, comienza en el 4 y cuenta 2.

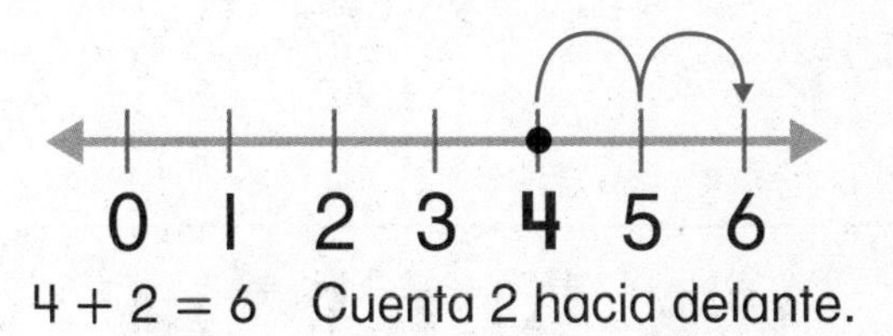

4 + 2 = 6 Cuenta 2 hacia delante.

cube A three-dimensional shape with 6 square faces.

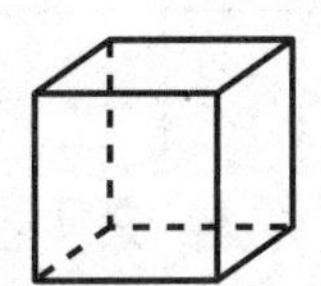

cubo Una figura tridimensional con 6 caras cuadradas.

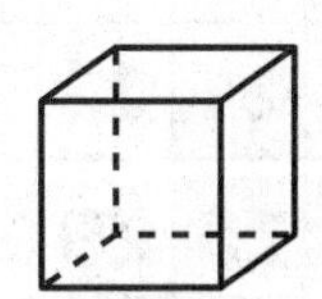

cylinder A three-dimensional shape that is shaped like a can.

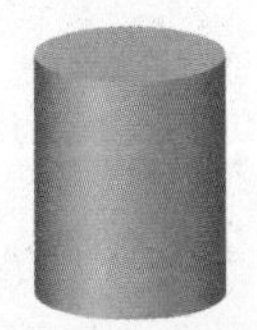
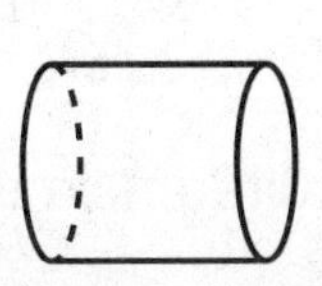

cilindro Una figura tridimensional que tiene la forma de una lata.

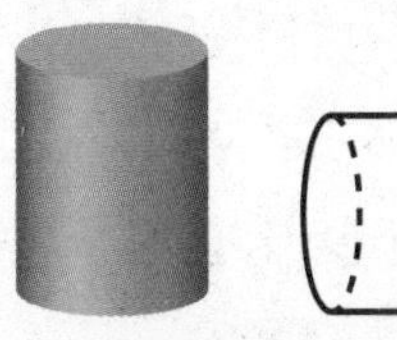

data Numbers or symbols collected to show information.

Name	Number of Pets
Mary	3
James	1
Alonzo	4

datos Números o símbolos que se recopilan para mostrar información.

Nombre	Número de mascotas
Maria	3
James	1
Alonzo	4

day

día

difference The answer to a subtraction problem.

$3 - 1 = 2$

The difference is 2.

diferencia Resultado de un problema de resta.

$3 - 1 = 2$

La diferencia es 2.

digital clock A clock that uses only numbers to show time.

reloj digital Reloj que usa solo números para mostrar la hora.

dime dime = 10¢ or 10 cents

head

tail

moneda de 10¢ moneda de diaz centavos = 10¢ o 10 centavos

cara

cruz

doubles (doubles plus 1, near doubles) Two addends that are the same number.

$2 + 2 = 4$

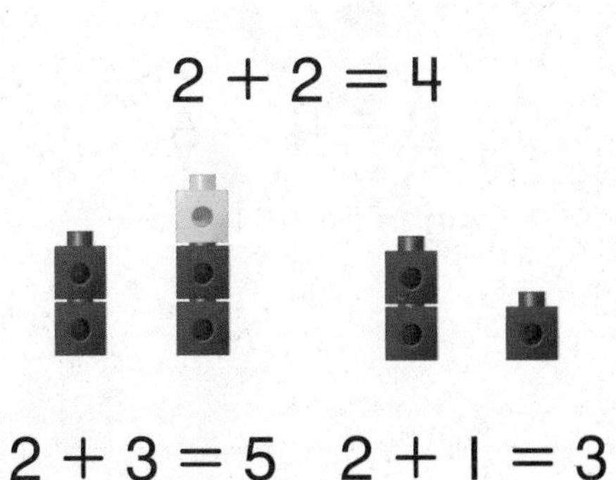

$2 + 3 = 5$ $2 + 1 = 3$

dobles (y dobles más 1, casi dobles) Dos sumandos que son el mismo número.

$2 + 2 = 4$

$2 + 3 = 5$ $2 + 1 = 3$

Ee

equal parts Each part is the same size.

A muffin cut in equal parts.

partes iguales Cada parte es del mismo tamaño.

Un panecillo cortado en partes iguales.

equal to =

6 = 6

6 is equal to 6.

igual a =

6 = 6

6 es igual a 6.

equals (=) Having the same value as or is the same as.

2 + 4 = 6

equals sign ↑

igual (=) Que tienen el mismo valor o son lo mismo.

2 + 4 = 6

signo igual ↑

face The flat part of a three-dimensional shape.

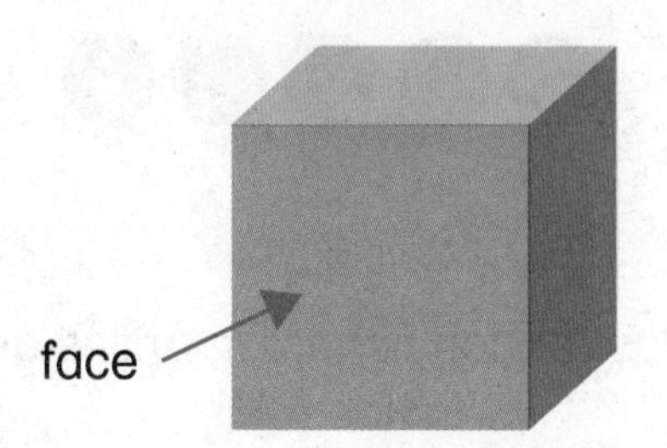

cara La parte plana de una figura tridimensional.

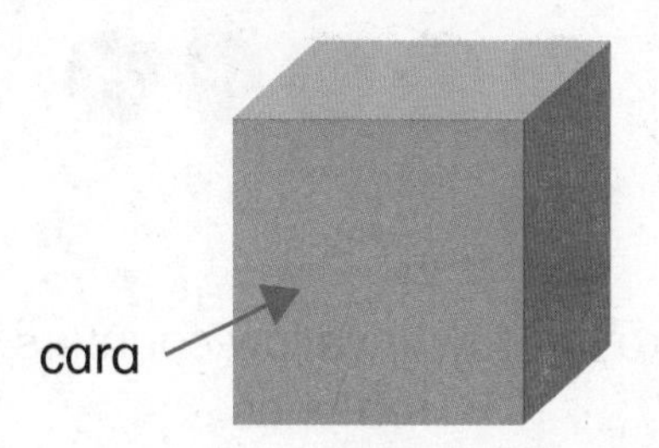

fact family Addition and subtraction sentences that use the same numbers. Sometimes called *related facts*.

$6 + 7 = 13$ $13 - 7 = 6$

$7 + 6 = 13$ $13 - 6 = 7$

familia de operaciones Enunciados de suma y resta que tienen los mismos números. Algunas veces se llaman *operaciones relacionadas*.

$6 + 7 = 13$ $13 - 7 = 6$

$7 + 6 = 13$ $13 - 6 = 7$

false Something that is not a fact. The opposite of true.

falso Algo que no es cierto. Lo opuesto de verdadero.

Ff

fewer/fewest The number or group with less.

There are fewer yellow counters than red ones.

menos/el menor El número o grupo con menos.

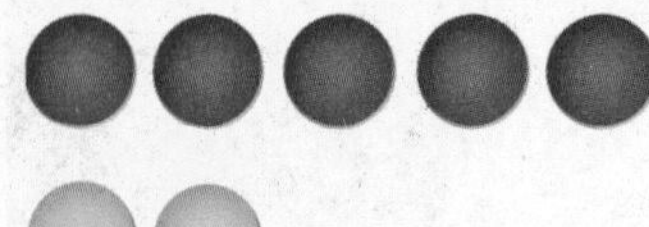

Hay menos fichas amarillas que fichas rojas.

fourths Four equal parts of a whole. Each part is a fourth, or a quarter of the whole.

cuartos Cuatro partes iguales de un todo. Cada parte es un cuarto, o la cuarta parte del todo.

Gg

graph A way to present data collected.

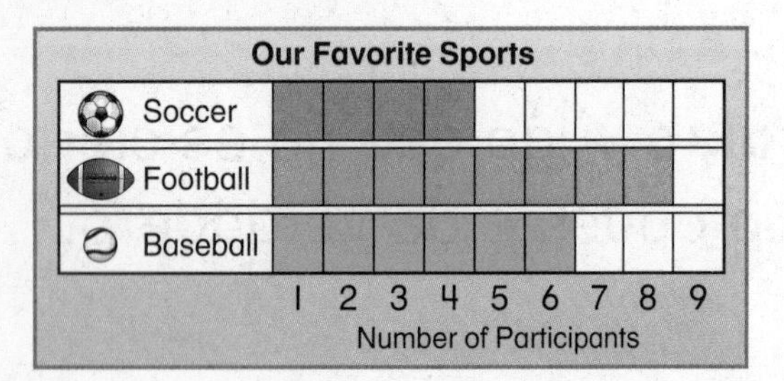

bar graph

gráfica Forma de presentar datos recopilados.

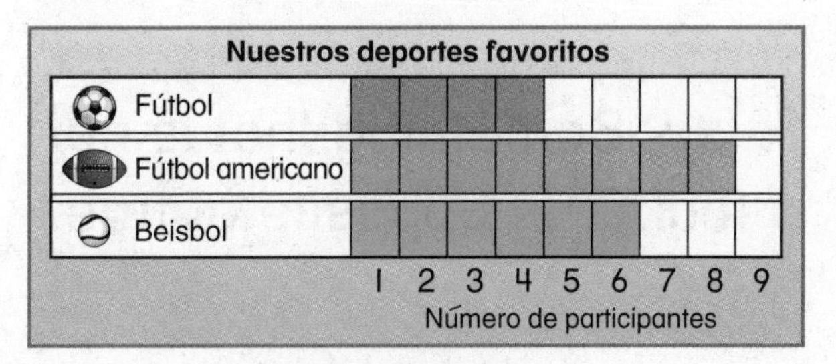

gráfica de barras

greater than (>)/greatest The number or group with more.

4 23 56

56 is the greatest.

mayor que (>)/el mayor El número o grupo con más cantidad.

4 23 56

56 es el mayor.

Hh

half hour (or half past) One half of an hour is 30 minutes. Sometimes called *half past* or *half past the hour.*

media hora (o y media) Media hora son 30 minutos. A veces se dice *hora y media.*

halves Two equal parts of a whole. Each part is a half of the whole.

mitades Dos partes iguales de un todo. Cada parte es la mitad de un todo.

heavy (heavier, heaviest) Weighs more.

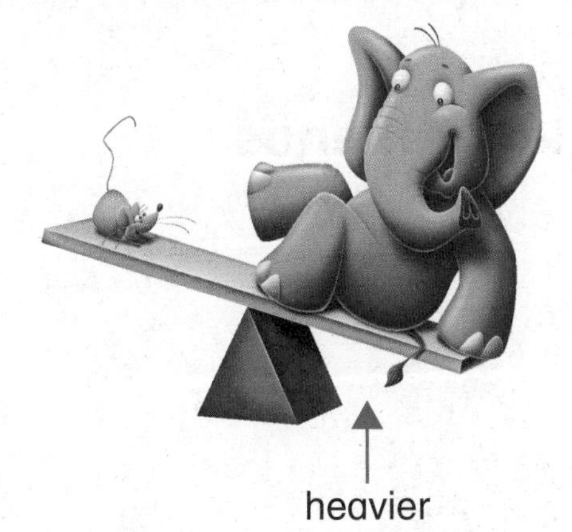

An elephant is heavier than a mouse.

pesado (más pesado, el más pesado) Pesa más.

Un elefante es más pesado (pesa más) que un ratón.

height

altura

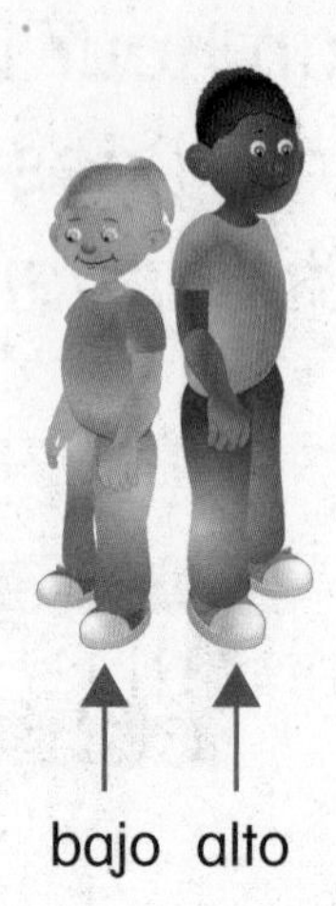

hexagon A two-dimensional shape that has six sides.

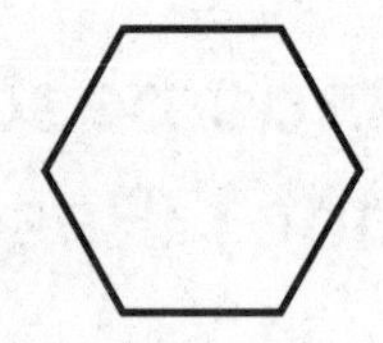

hexágono Figura bidimensional que tiene seis lados.

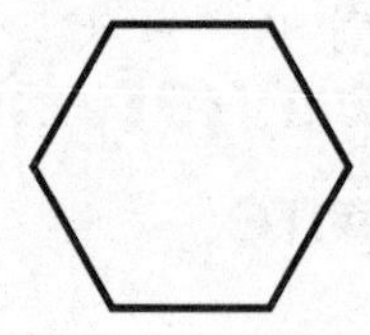

holds less/least

The glass holds less than the pitcher.

contener menos

El vaso contiene menos que la jarra.

holds more/most

The pitcher holds more than the glass.

contener más

La jarra contiene más que el vaso.

hour A unit of time.

1 hour = 60 minutes

hora Unidad de tiempo.

1 hora = 60 minutos

hour hand The hand on a clock that tells the hour. It is the shorter hand.

manecilla horaria Manecilla del reloj que indica la hora. Es la manecilla más corta.

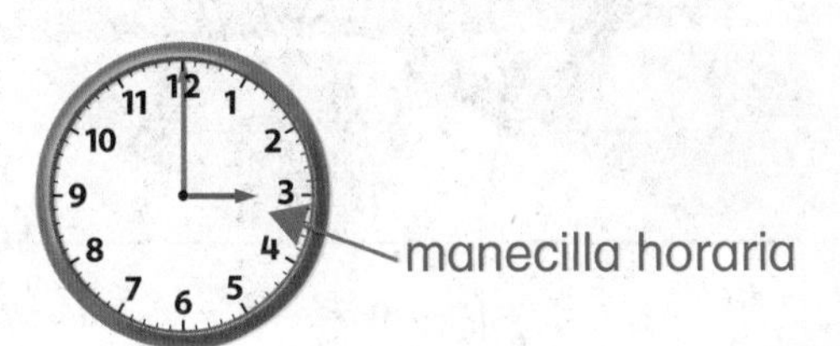

Hh

hundreds The numbers in the range of 100-999. It is the place value of a number.

centenas Los números en el rango del 100 al 999. Es el valor posicional de un número.

inverse Operations that undo each other.

Addition and subtraction are inverse or opposite operations.

operaciones inversas Operaciones que se anulan entre sí.

La suma y la resta son operaciones inversas u opuestas.

length

longitud

less than (<)/least The number or group with fewer.

4 23 56

4 is the least.

light (lighter, lightest) Weighs less.

lighter

The mouse is lighter than the elephant.

long (longer, longest) A way to compare the lengths of two objects.

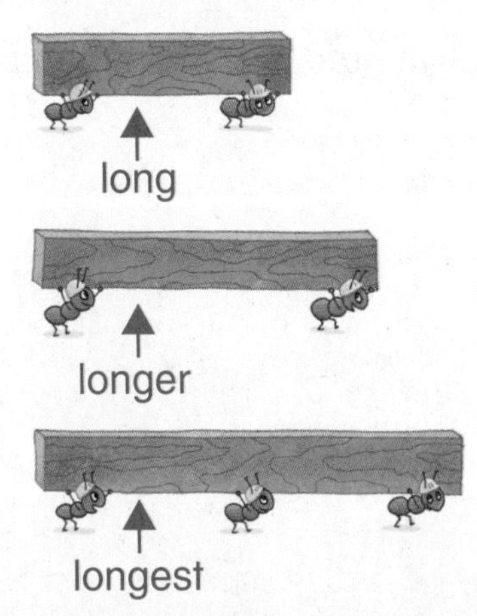

menor que (<)/el menor El número o grupo con menos cantidad.

4 23 56

4 es el menor.

liviano (más liviano, el más liviano) Pesa menos.

más liviano

El ratón es más liviano (pesa menos) que el elefante.

largo (más largo, el más largo) Forma de comparar la longitud de dos objetos.

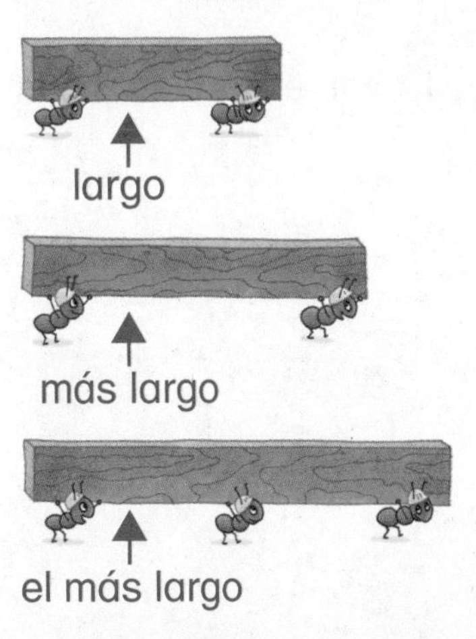

Mm

mass The amount of matter in an object. The mass of an object never changes.

masa Cantidad de materia en un objeto. La masa de un cuerpo nunca cambia.

measure To find the length, height, weight or capacity using standard or nonstandard units.

medir Hallar la longitud, altura, peso o capacidad mediante unidades estándar o no estándar.

minus (–) The sign used to show subtraction.

$5 - 2 = 3$

minus sign

menos (–) Signo que indica resta.

$5 - 2 = 3$

signo menos

minute (min) A unit to measure time.

1 minute = 60 seconds

minuto (min) Unidad que se usa para medir el tiempo.

1 minuto = 60 segundos

minute hand The longer hand on a clock that tells the minutes.

minutero La manecilla más larga del reloj. Indica los minutos.

missing addend

$9 + \_\_\_\_ = 16$

The missing addend is 7.

sumando desconocido

$9 + \_\_\_\_ = 16$

El sumando desconocido es 7.

month

month

April

Sunday	Monday	Tuesday	Wednesday	Thursday	Friday	Saturday
		1	2	3	4	5
6	7	8	9	10	11	12
13	14	15	16	17	18	19
20	21	22	23	24	25	26
27	28	29	30			

mes

mes

abril

domingo	lunes	martes	miércoles	jueves	viernes	sábado
		1	2	3	4	5
6	7	8	9	10	11	12
13	14	15	16	17	18	19
20	21	22	23	24	25	26
27	28	29	30			

Mm

more

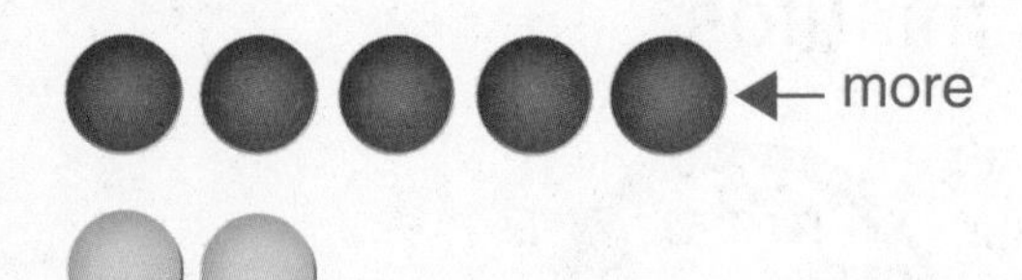

más

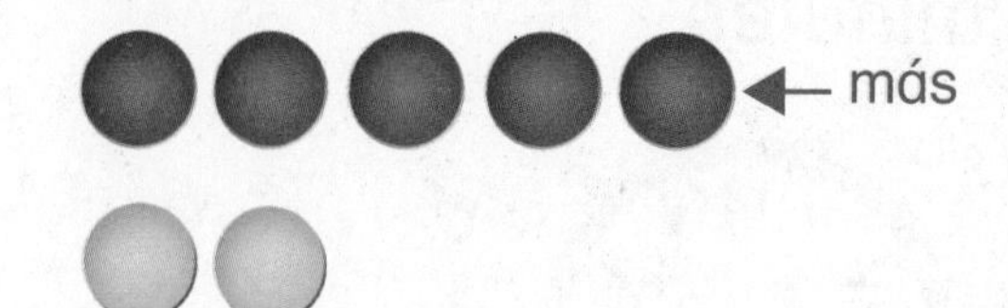

Nn

nickel nickel = 5¢ or 5 cents

head tail

moneda de 5¢ moneda de cinco centavos = 5¢ o 5 centavos

cara cruz

number Tells how many.
1, 2, 3, 4, 5, 6, 7, 8, 9, 10 ...

There are 3 chicks.

número Dice cuántos hay.
1, 2, 3, 4, 5, 6, 7, 8, 9, 10 ...

Hay tres pollitos.

number line A line with number labels.

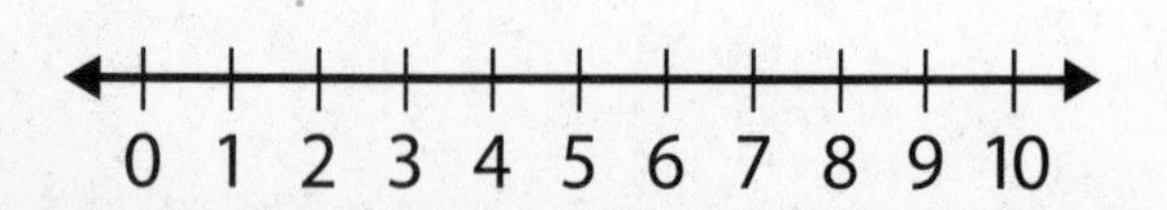

recta numérica Recta con marcas de números.

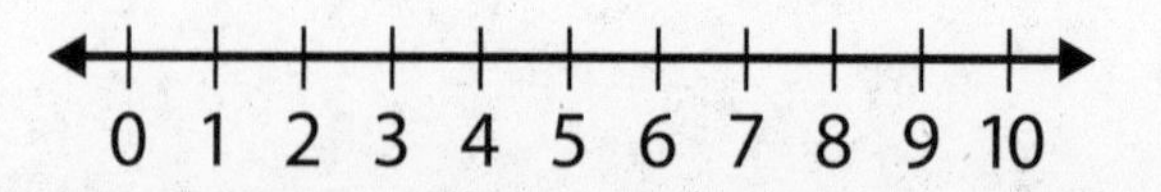

Oo

o'clock At the beginning of the hour.

It is 3 o'clock.

en punto Al comienzo de la hora.

Son las 3 en punto.

ones The numbers in the range of 0–9. It is the place value of a number.

unidades Los números en el rango de 0 a 9. Es el valor posicional de un número.

order

1, 3, 6, 7, 9

These numbers are in order from least to greatest.

orden

1, 3, 6, 7, 9

Estos números están en orden del menor al mayor.

ordinal number

first second third

númeral ordinal

primero segundo tercero

Pp

part One of the parts joined when adding.

Part	Part
2	2
Whole	

pattern An order that a set of objects or numbers follows over and over.

A, A, B, A, A, B, A, A, B

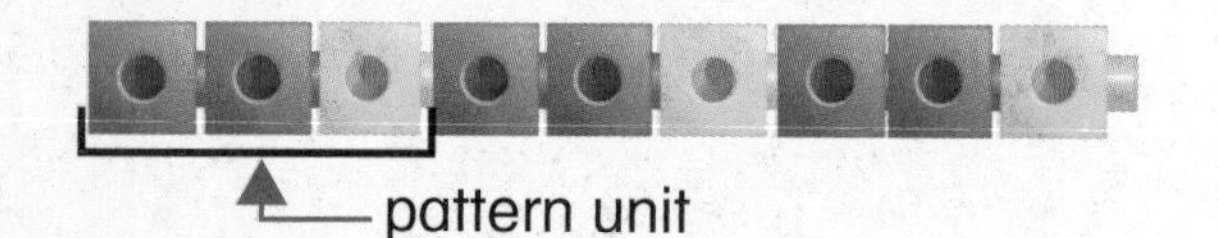

penny penny = 1¢ or 1 cent

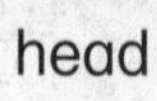

head tail

parte Una de las partes que se juntan al sumar.

Parte	Parte
2	2
El total	

patrón Orden que sigue continuamente un conjunto de objetos o números.

A, A, B, A, A, B, A, A, B

moneda de 1¢ moneda de un centavo = 1¢ o 1 centavo

cara cruz

picture graph A graph that has different pictures to show information collected.

gráfica con imágenes Gráfica que tiene diferentes imágenes para ilustrar la información recopilada.

place value The value given to a digit by its place in a number.

53

5 is in the tens place.
3 is in the ones place.

valor posicional Valor de un *dígito* según el lugar en el número.

53

5 está en el lugar de las decenas.
3 está en el lugar de las unidades.

plus (+) The sign used to show addition.

$4 + 5 = 9$

plus sign

más (+) Símbolo para mostrar la suma.

$4 + 5 = 9$

signo más

Pp

position Tells where an object is.

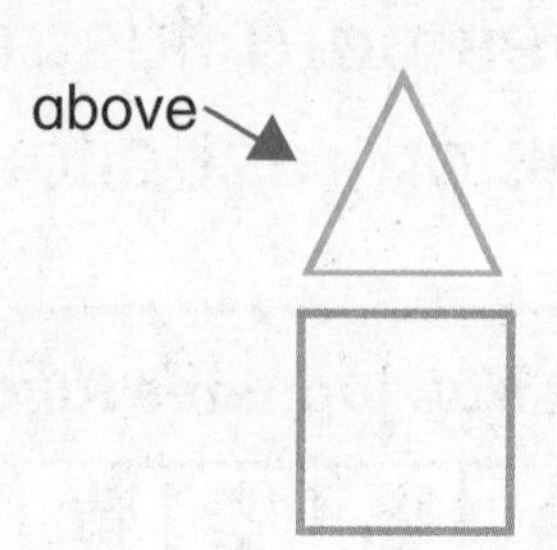

posición Indica dónde está un objeto.

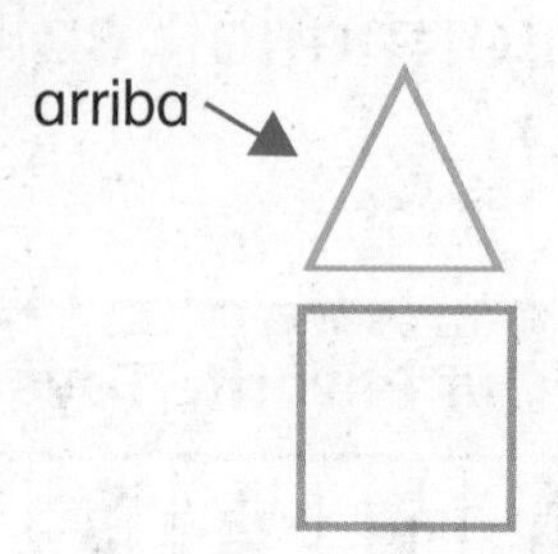

rectangle A shape with four sides and four corners.

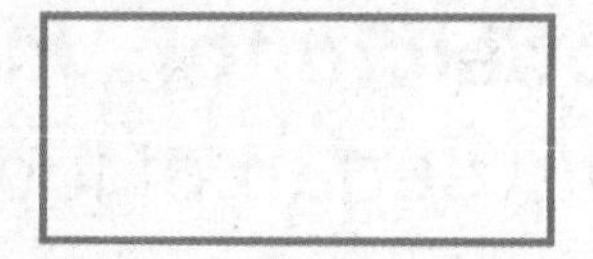

rectángulo Figura con cuatro lados y cuatro esquinas.

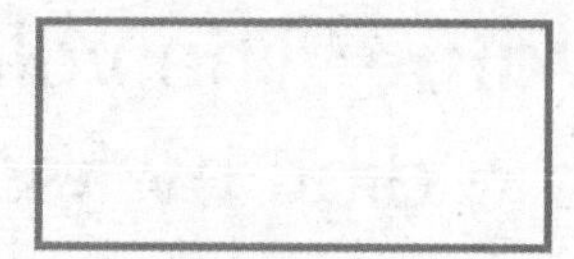

rectangular prism A three-dimensional shape with 6 faces that are rectangles.

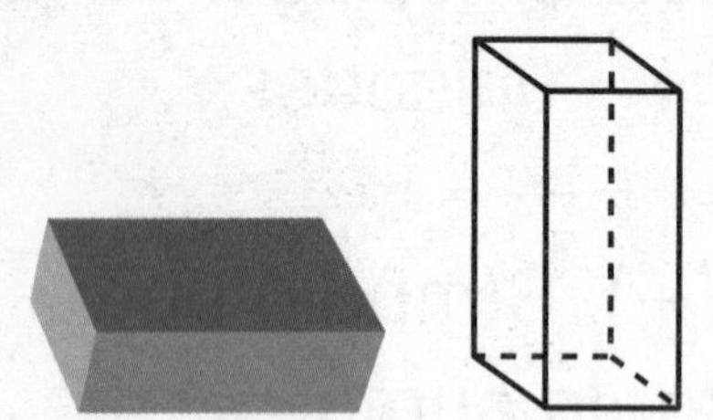

prisma rectangular Una figura tridimensional con 6 caras que son rectángulos.

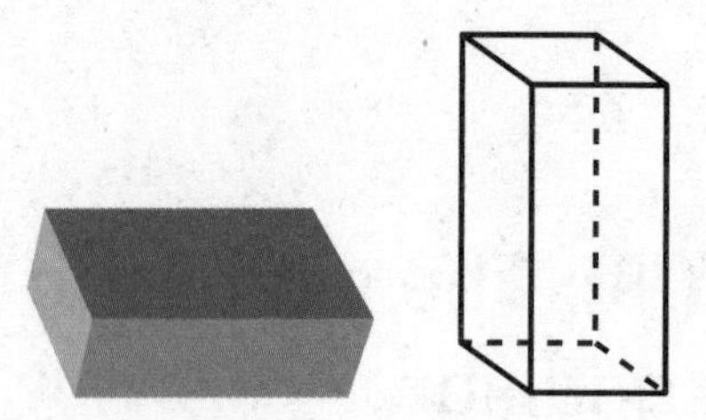

regroup To take apart a number to write it in a new way.

1 ten + 2 ones becomes 12 ones.

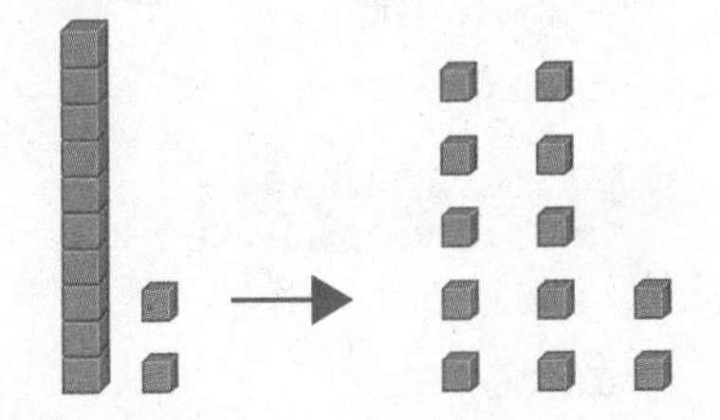

reagrupar Separar un número para escribirlo en una nueva forma.

1 decena + 2 unidades se convierten en 12 unidades.

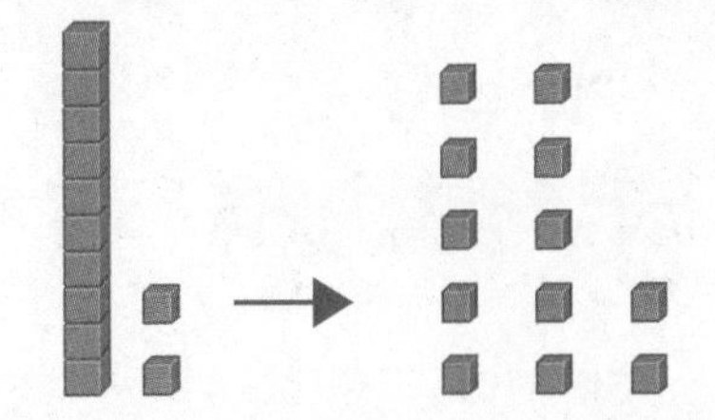

related fact(s) Basic facts using the same numbers. Sometimes called a *fact family*.

$4 + 1 = 5$ $\quad 5 - 4 = 1$
$1 + 4 = 5$ $\quad 5 - 1 = 4$

operaciones relacionadas Operaciones básicas en las cuales se usan los mismos números. También se llaman *familias de operaciones.*

$4 + 1 = 5$ $\quad 5 - 4 = 1$
$1 + 4 = 5$ $\quad 5 - 1 = 4$

repeating pattern

patrón repetitivo

Ss

short (shorter, shortest) To compare length or height of two (or more) objects.

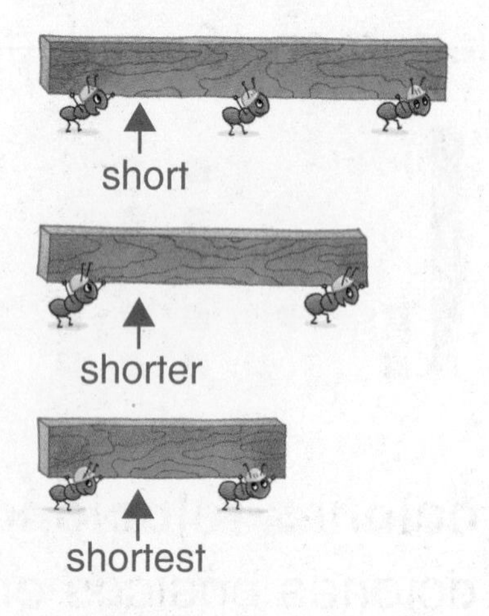

corto (más corto, el más corto) Comparar la longitud o la altura de dos (o más) objetos.

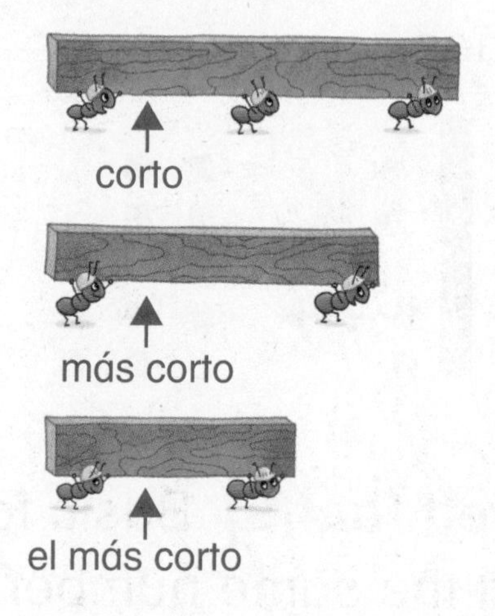

side

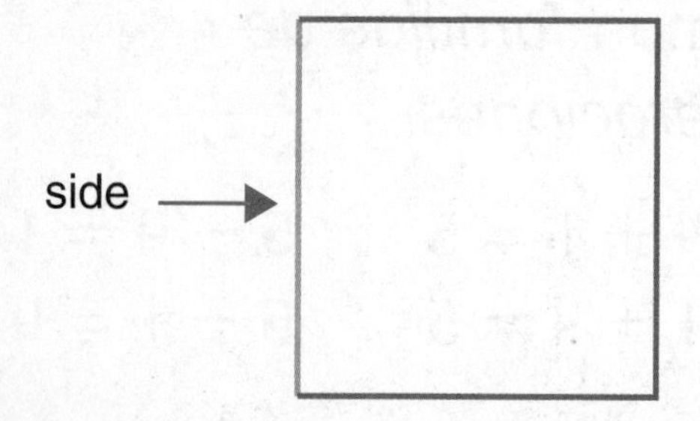

lado

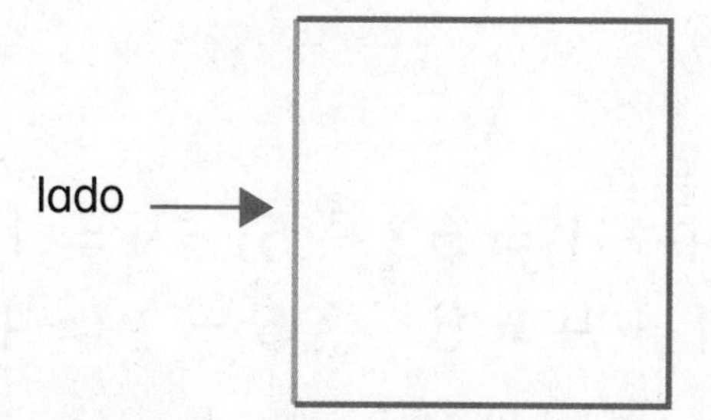

sort To group together like items.

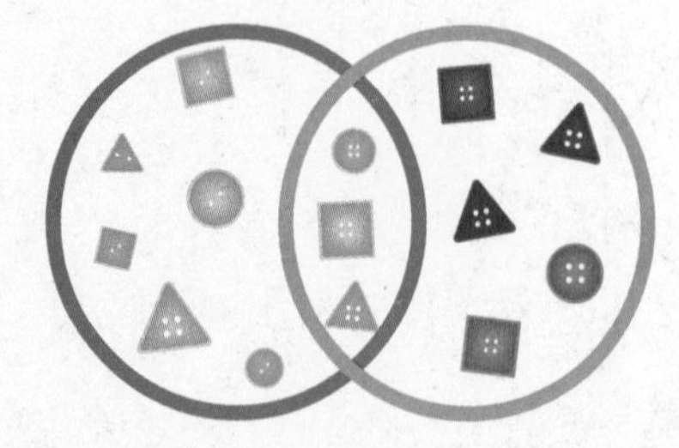

clasificar Agrupar elementos con iguales características.

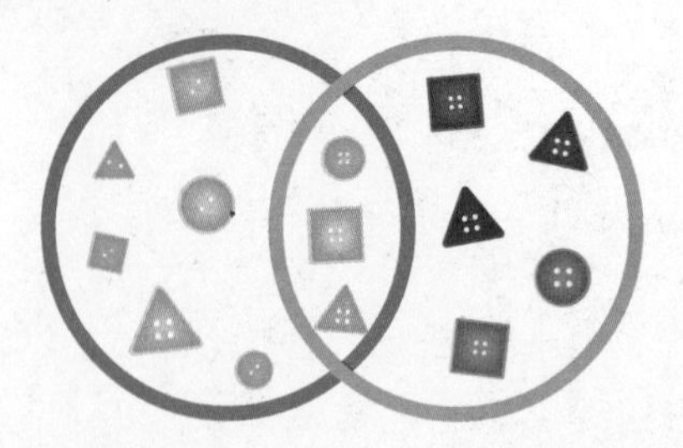

sphere A solid shape that has the shape of a round ball.

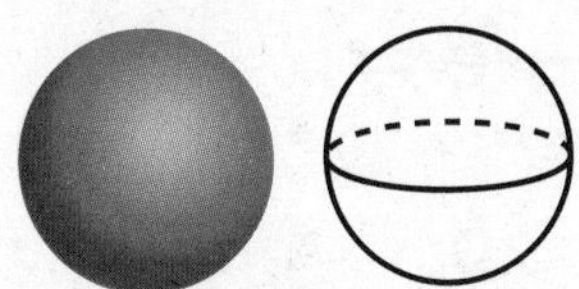

esfera Un sólido con la forma de una pelota redonda.

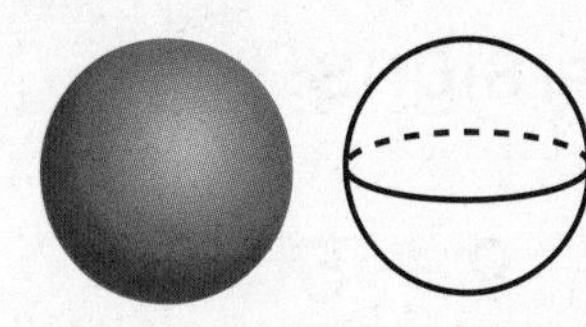

square A rectangle that has four equal sides.

cuadrado Rectángulo que tiene cuatro lados iguales.

subtract (subtracting, subtraction) To take away, take apart, separate, or find the difference between two sets. The opposite of addition.

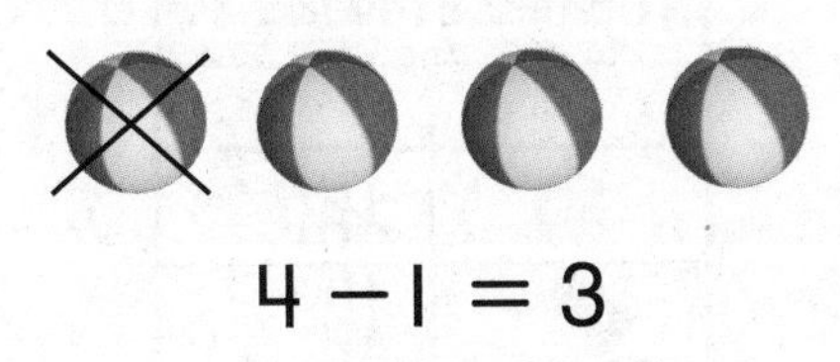

$4 - 1 = 3$

restar (resta, sustracción) Eliminar, quitar, separar o hallar la diferencia entre dos conjuntos. Lo opuesto de la suma.

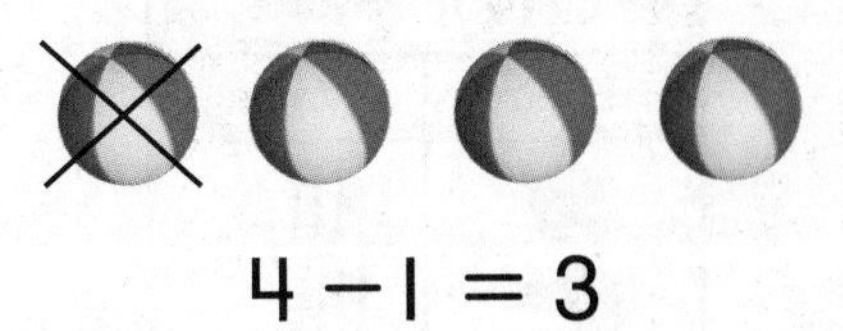

$4 - 1 = 3$

Ss

subtraction number sentence An expression using numbers and the − and = signs.

$9 - 5 = 4$

enunciado numérico de resta Expresión en la cual se usan números con los signos − e =.

$9 - 5 = 4$

sum The answer to an addition problem.

$2 + 4 = 6$

↑ sum

suma Resultado de la operación de sumar.

$2 + 4 = 6$

↑ suma

survey To collect data by asking people the same question.

Favorite Foods	
Food	Votes
	卌
	|||
	卌 |||

This survey shows favorite foods.

encuesta Recopilación de datos haciendo las mismas preguntas a un grupo de personas.

Comidas favoritas	
Comida	Votos
	卌
	|||
	卌 |||

Esta encuesta muestra las comidas favoritas.

Tt

tall (taller, tallest)

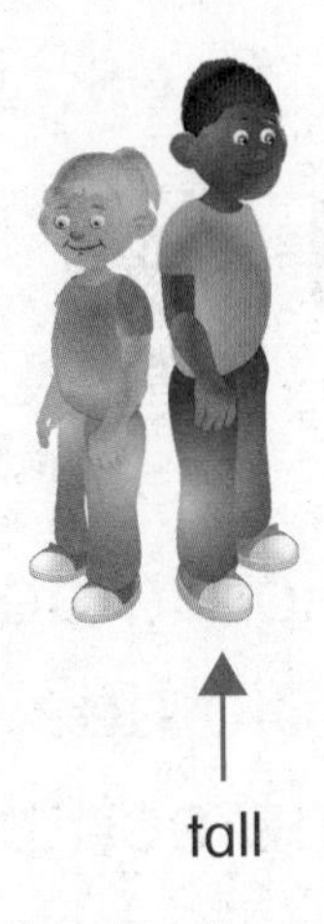

alto (más alto, el más alto)

tally chart A way to show data collected using tally marks.

Favorite Foods	
Food	Votes
apple	~~IIII~~
fish	III
sandwich	~~IIII~~ III

tabla de conteo Forma de mostrar los datos recopilados utilizando marcas de conteo.

Comidas favoritas	
Comida	Votos
manzana	~~IIII~~
pescado	III
sándwich	~~IIII~~ III

tens The numbers in the range 10–99. It is the place value of a number.

53

5 is in the tens place.
3 is in the ones place.

decenas Los números en el rango del 10 al 99. Es el valor posicional de un número.

53

5 está en el lugar de las decenas.
3 está en el lugar de las unidades.

Tt

three-dimensional shape
A solid shape.

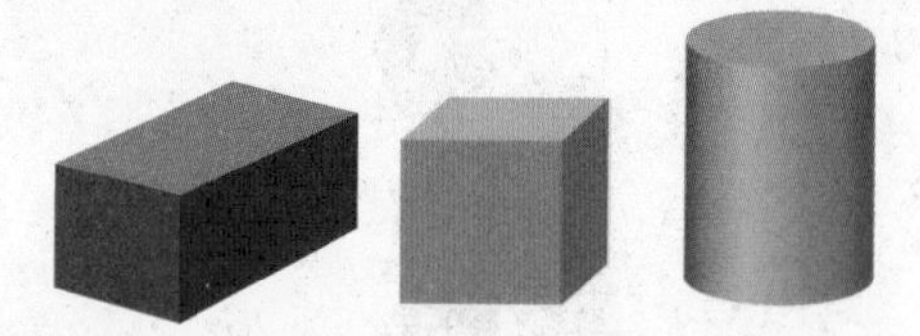

figura tridimensional
Un sólido.

trapezoid A four-sided plane shape with only two opposite sides that are parallel.

trapecio Figura de cuatro lados con solo dos lados opuestos que son paralelos.

triangle A shape with three sides.

triángulo Figura con tres lados.

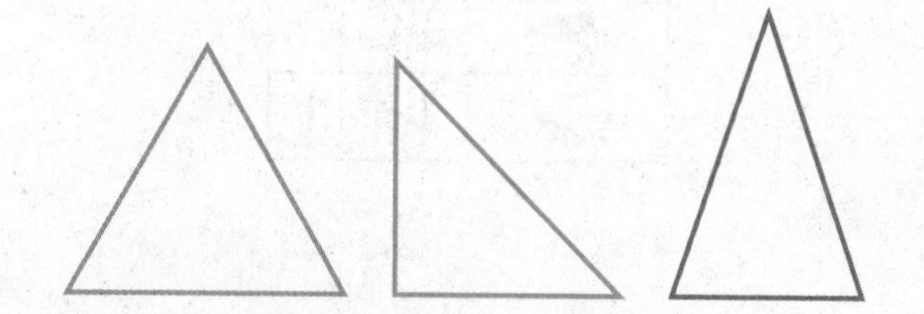

true Something that is a fact. The opposite of false.

verdadero Algo que es cierto. Lo opuesto de falso.

two-dimensional shape The outline of a shape such as a triangle, square, or rectangle.

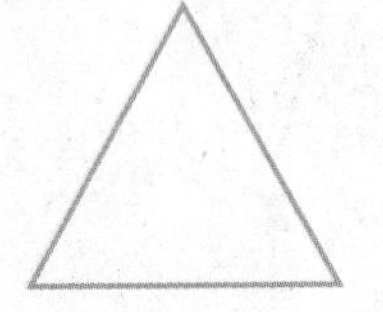 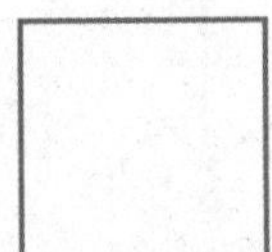 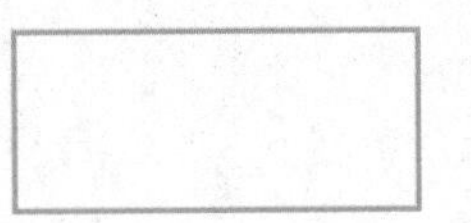

figura bidimensional Contorno de una figura como un triángulo, o un cuadrado rectángulo.

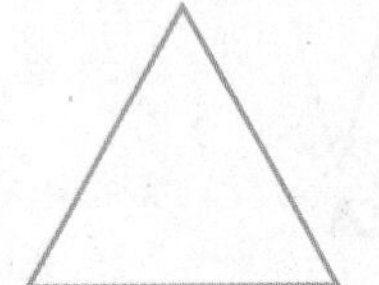 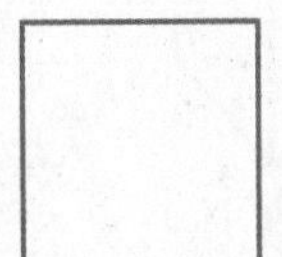 

Uu

unit An object used to measure.

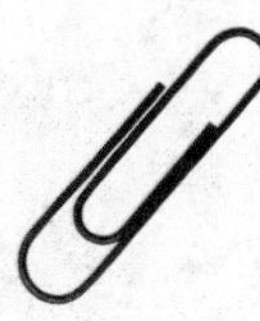 

unidad Objeto que se usa para medir.

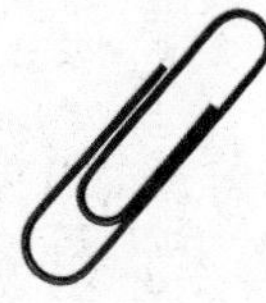 

Vv

Venn diagram A drawing that uses circles to sort and show data.

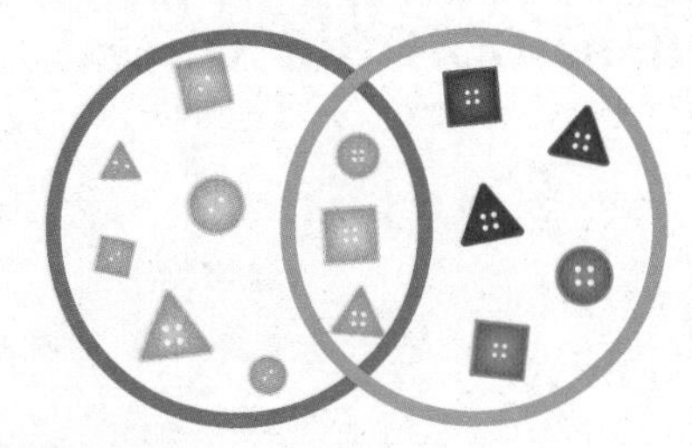

diagrama de Venn Dibujo que tiene círculos para clasificar y mostrar datos.

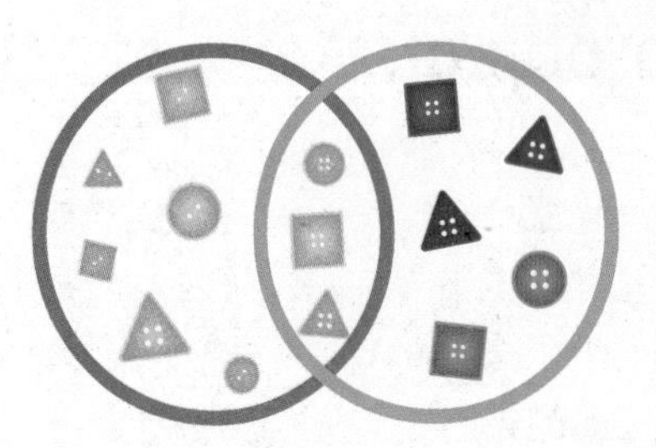

Vv

vertex

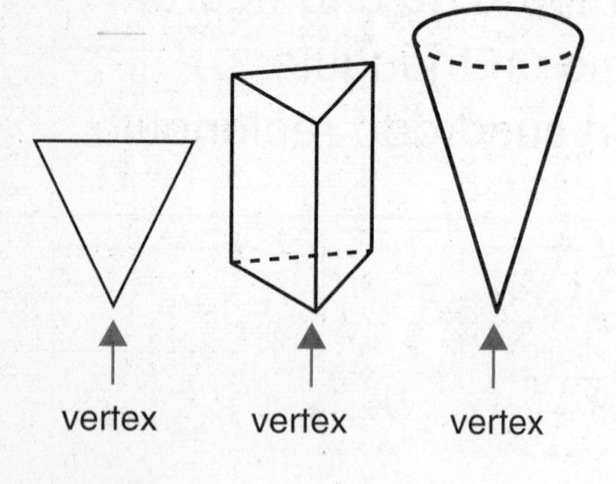

vértice

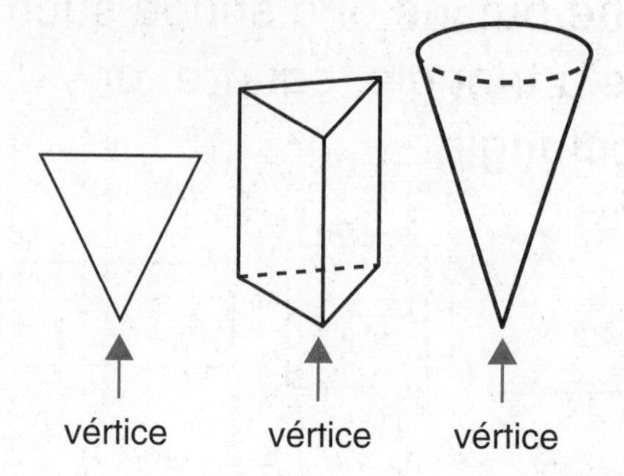

Ww

weight

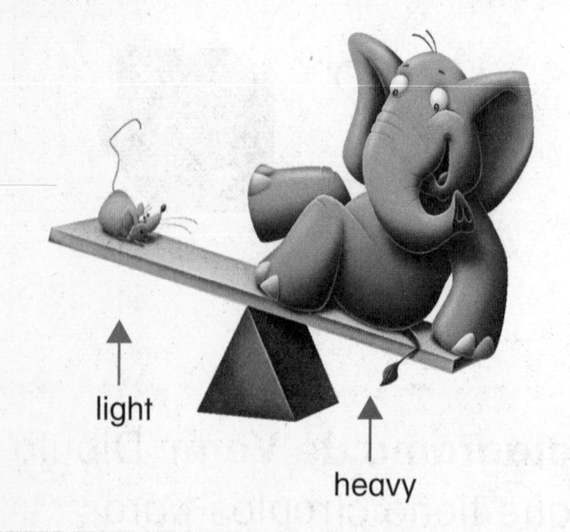

peso

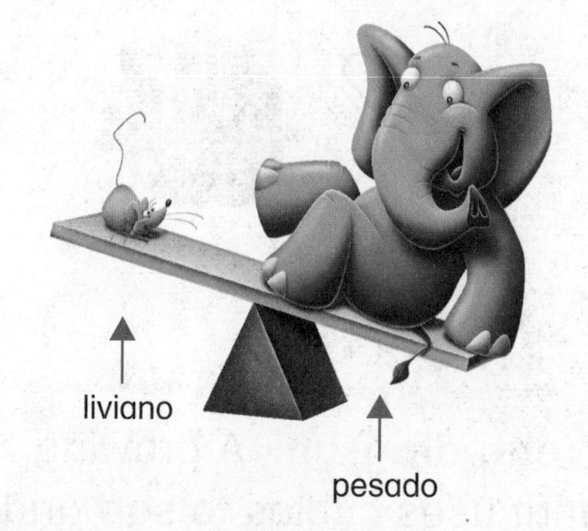

whole The entire amount of an object.

el todo La cantidad total o el objeto completo.

Yy

year

January

S	M	T	W	T	F	S
						1
2	3	4	5	6	7	8
9	10	11	12	13	14	15
16	17	18	19	20	21	22
23	24	25	26	27	28	29
30	31					

February

S	M	T	W	T	F	S
		1	2	3	4	5
6	7	8	9	10	11	12
13	14	15	16	17	18	19
20	21	22	23	24	25	26
27	28					

March

S	M	T	W	T	F	S
		1	2	3	4	5
6	7	8	9	10	11	12
13	14	15	16	17	18	19
20	21	22	23	24	25	26
27	28	29	30	31		

April

S	M	T	W	T	F	S
					1	2
3	4	5	6	7	8	9
10	11	12	13	14	15	16
17	18	19	20	21	22	23
24	25	26	27	28	29	30

May

S	M	T	W	T	F	S
1	2	3	4	5	6	7
8	9	10	11	12	13	14
15	16	17	18	19	20	21
22	23	24	25	26	27	28
29	30	31				

June

S	M	T	W	T	F	S
			1	2	3	4
5	6	7	8	9	10	11
12	13	14	15	16	17	18
19	20	21	22	23	24	25
26	27	28	29	30		

July

S	M	T	W	T	F	S
					1	2
3	4	5	6	7	8	9
10	11	12	13	14	15	16
17	18	19	20	21	22	23
24	25	26	27	28	29	30
31						

August

S	M	T	W	T	F	S
	1	2	3	4	5	6
7	8	9	10	11	12	13
14	15	16	17	18	19	20
21	22	23	24	25	26	27
28	29	30	31			

September

S	M	T	W	T	F	S
				1	2	3
4	5	6	7	8	9	10
11	12	13	14	15	16	17
18	19	20	21	22	23	24
25	26	27	28	29	30	

October

S	M	T	W	T	F	S
						1
2	3	4	5	6	7	8
9	10	11	12	13	14	15
16	17	18	19	20	21	22
23	24	25	26	27	28	29
30	31					

November

S	M	T	W	T	F	S
		1	2	3	4	5
6	7	8	9	10	11	12
13	14	15	16	17	18	19
20	21	22	23	24	25	26
27	28	29	30			

December

S	M	T	W	T	F	S
				1	2	3
4	5	6	7	8	9	10
11	12	13	14	15	16	17
18	19	20	21	22	23	24
25	26	27	28	29	30	31

año

enero

d	l	m	m	j	v	s
						1
2	3	4	5	6	7	8
9	10	11	12	13	14	15
16	17	18	19	20	21	22
23	24	25	26	27	28	29
30	31					

febrero

d	l	m	m	j	v	s
		1	2	3	4	5
6	7	8	9	10	11	12
13	14	15	16	17	18	19
20	21	22	23	24	25	26
27	28					

marzo

d	l	m	m	j	v	s
		1	2	3	4	5
6	7	8	9	10	11	12
13	14	15	16	17	18	19
20	21	22	23	24	25	26
27	28	29	30	31		

abril

d	l	m	m	j	v	s
					1	2
3	4	5	6	7	8	9
10	11	12	13	14	15	16
17	18	19	20	21	22	23
24	25	26	27	28	29	30

mayo

d	l	m	m	j	v	s
1	2	3	4	5	6	7
8	9	10	11	12	13	14
15	16	17	18	19	20	21
22	23	24	25	26	27	28
29	30	31				

junio

d	l	m	m	j	v	s
			1	2	3	4
5	6	7	8	9	10	11
12	13	14	15	16	17	18
19	20	21	22	23	24	25
26	27	28	29	30		

julio

d	l	m	m	j	v	s
					1	2
3	4	5	6	7	8	9
10	11	12	13	14	15	16
17	18	19	20	21	22	23
24	25	26	27	28	29	30
31						

agosto

d	l	m	m	j	v	s
	1	2	3	4	5	6
7	8	9	10	11	12	13
14	15	16	17	18	19	20
21	22	23	24	25	26	27
28	29	30	31			

septiembre

d	l	m	m	j	v	s
				1	2	3
4	5	6	7	8	9	10
11	12	13	14	15	16	17
18	19	20	21	22	23	24
25	26	27	28	29	30	

octubre

d	l	m	m	j	v	s
						1
2	3	4	5	6	7	8
9	10	11	12	13	14	15
16	17	18	19	20	21	22
23	24	25	26	27	28	29
30	31					

noviembre

d	l	m	m	j	v	s
		1	2	3	4	5
6	7	8	9	10	11	12
13	14	15	16	17	18	19
20	21	22	23	24	25	26
27	28	29	30			

diciembre

d	l	m	m	j	v	s
				1	2	3
4	5	6	7	8	9	10
11	12	13	14	15	16	17
18	19	20	21	22	23	24
25	26	27	28	29	30	31

Zz

zero The number zero equals none or nothing.

cero El número cero es igual a nada o ninguno.

Name ....

Work Mat 4: Number Lines

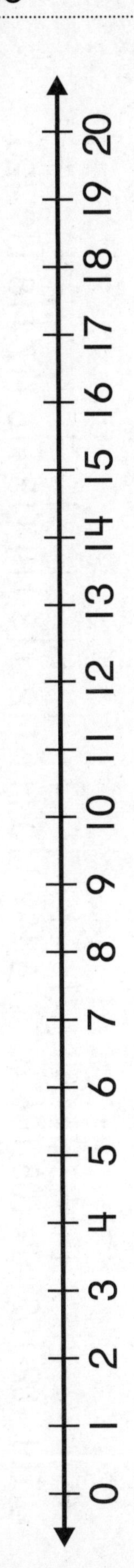

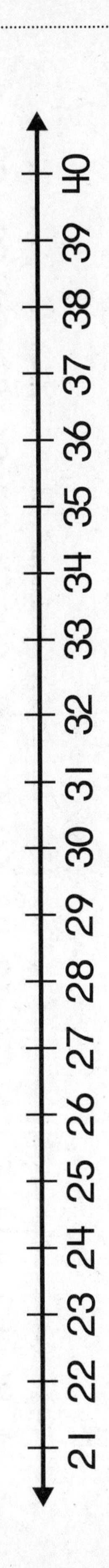

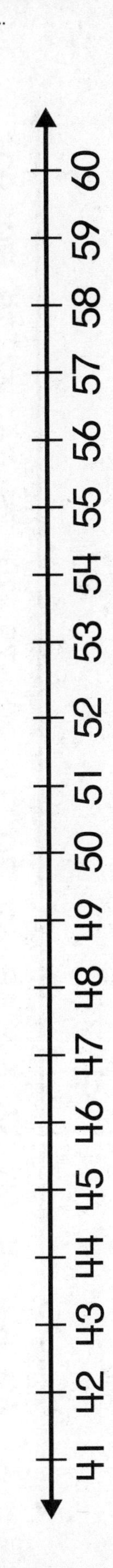

Work Mat 5: Number Lines

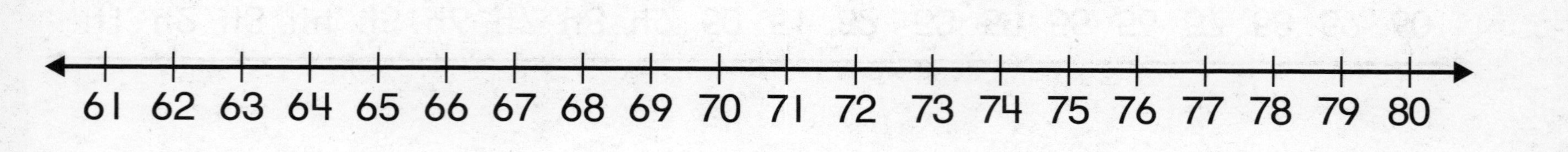

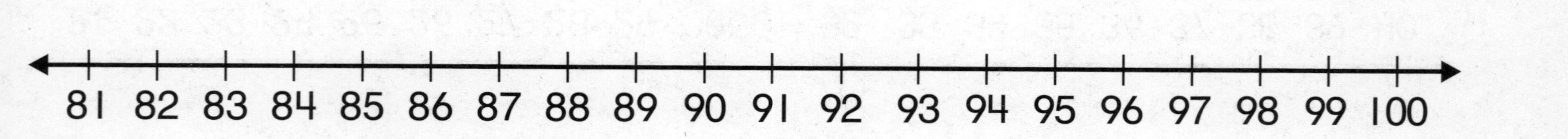

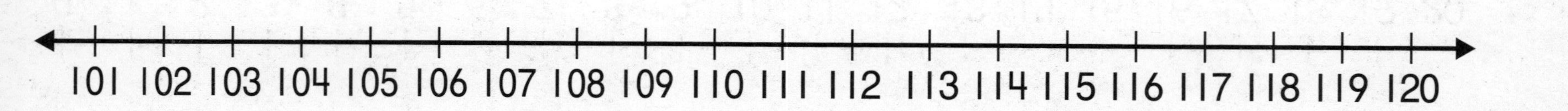

Work Mat 6: Grid

Work Mat 7: Tens and Ones Chart

Tens	Ones